# THE MAKING OF BRITISH ANTHROPOLOGY, 1813–1871

# Science and Culture in the Nineteenth Century

*Series Editor:* *Bernard Lightman*

## Titles in this Series

1 Styles of Reasoning in the British Life Sciences: Shared Assumptions, 1820–1858
*James Elwick*

2 Recreating Newton: Newtonian Biography and the Making of Nineteenth-Century History of Science
*Rebekah Higgitt*

3 The Transit of Venus Enterprise in Victorian Britain
*Jessica Ratcliff*

4 Science and Eccentricity: Collecting, Writing and Performing Science for Early Nineteenth-Century Audiences
*Victoria Carroll*

5 Typhoid in Uppingham: Analysis of a Victorian Town and School in Crisis, 1875–1877
*Nigel Richardson*

6 Medicine and Modernism: A Biography of Sir Henry Head
*L. S. Jacyna*

7 Domesticating Electricity: Expertise, Uncertainty and Gender, 1880–1914
*Graeme Gooday*

8 James Watt, Chemist: Understanding the Origins of the Steam Age
*David Philip Miller*

9 Natural History Societies and Civic Culture in Victorian Scotland
*Diarmid A. Finnegan*

10 Communities of Science in Nineteenth-Century Ireland
*Juliana Adelman*

11 Regionalizing Science: Placing Knowledges in Victorian England
*Simon Naylor*

12 The Science of History in Victorian Britain: Making the Past Speak
*Ian Hesketh*

13 Communicating Physics: The Production, Circulation and Appropriation of Ganot's Textbooks in France and England, 1851–1887
*Josep Simon*

14 The British Arboretum: Trees, Science and Culture in the Nineteenth Century
*Paul A. Elliott, Charles Watkins and Stephen Daniels*

15 Vision, Science and Literature, 1870–1920: Ocular Horizons
*Martin Willis*

16 Popular Exhibitions, Science and Showmanship, 1840–1910
*Joe Kember, John Plunkett and Jill A. Sullivan (eds)*

17 Free Will and the Human Sciences in Britain, 1870–1910
*Roger Smith*

## Forthcoming Titles

Science and Societies in Frankfurt am Main
*Ayako Sakurai*

Brewing Science, Technology and Print, 1700–1880
*James Sumner*

The Making of Modern Anthrax, 1875–1920: Uniting Local, National and Global Histories of Disease
*James F. Stark*

The Medical Trade Catalogue in Britain, 1870–1914
*Claire L. Jones*

Uncommon Contexts: Encounters between Science and Literature, 1800–1914
*Ben Marsden, Hazel Hutchison and Ralph O'Connor (eds)*

Astronomy in India, 1784–1876
*Joydeep Sen*

The Age of Scientific Naturalism: Tyndall and his Contemporaries
*Bernard Lightman and Michael S. Reidy (eds)*

# THE MAKING OF BRITISH ANTHROPOLOGY, 1813–1871

BY

Efram Sera-Shriar

**Published by the University of Pittsburgh Press, Pittsburgh, Pa., 15260**

This paperback edition, 2020


Cataloging-in-Publication is available from the British Library

ISBN 13: 978-0-8229-6648-7
ISBN 10: 0-8229-6648-4

# CONTENTS

# ACKNOWLEDGEMENTS

My thanks extend to a large number of people who supported me throughout the writing of this book. The foundation of this text is based on my doctoral thesis, which I completed in 2011 at the University of Leeds. I would like to thank my doctoral supervisors, Gregory Radick and Jonathan Topham, for their guidance during my time at Leeds. Without their support this project would not have been possible. I must also recognize Graeme Gooday for his practical advice over the past six years. The faculty and postgraduate community at Leeds were extremely supportive, and I would like to thank them all for putting up with me during my most stressful moments of writing up parts of this work. My gratitude also extends to the Centre for History and Philosophy of Science at Leeds for providing me with funding during the first three years of my doctoral research. The John Tyndall Correspondence Project has been equally accommodating by helping me through the early part of my career.

My thanks also extend to the staff librarians at the University Library in Cambridge, the Wellcome Library in London, the Brotherton Library at the University of Leeds and the Scott Library at York University. I also want to thank the archivist Sarah Walpole at the Royal Anthropological Institute in London for providing me with access to their important collections and a workspace. I am also grateful to Phillip Grover and Christopher Morton at the Pitt Rivers Museum in Oxford for letting me view their photographic and manuscript collections. I would like to thank the members of the Science and Technology Studies Programme at York University in Canada for their support. In particular, Bernard Lightman and James Elwick deserve a special acknowledgement for helping me settle into postdoctoral life. I am also grateful to the Social Sciences and Humanities Research Council of Canada for providing me with a postdoctoral fellowship.

Over the past decade I have received tremendous support from my family and friends around the world. I would like to thank my wife, Nanna Kaalund, and my best friend, Geoffrey Belknap, for all the support they have given me. Thank you to Sujit Sivasundaram for being my external examiner and for his invaluable suggestions. I would also like to give special thanks to Sadiah Qureshi, Michael Reidy, Simon Schaffer, Peter Kjaergaard, Jonathan Hodge, Trevor Levere, Nicholas Jardine, Peter Mandler, James Delbourgo and Henrika Kuklick for all their advice and support during my research. I am grateful to my two anonymous referees for their detailed reports and to the editorial team at Pickering & Chatto for helping to bring this book to press. Finally, I would like to thank Geoffrey Cantor, without whom I would never have applied to work at Leeds and write this book. Your guidance over the past few years has meant so much to me.

# LIST OF FIGURES

# INTRODUCTION

> Reason and logic is itself a science, but like other sciences, it began as an art which man practised without stopping to ask himself why or how.[1]
>
> Edward Burnett Tylor, 1881

## Rethinking the Significance of the Field

In the introduction to his book *Argonauts of the Western Pacific* (1922), the anthropologist Bronislaw Malinowski (1884–1942) argued that it was important for researchers interested in human diversity to discuss the methods they used in the field when collecting ethnographic materials. It was only by outlining in detail these rigorous practices that researchers would be able to demonstrate the scientific standards of the discipline. Malinowski wrote, 'it will be well to give a description of the methods used in the collecting of ethnographic materials. The results of scientific research in any branch of learning ought to be presented in a manner absolutely candid and above board'.[2] He continued by linking transformations in anthropological techniques to the larger changes occurring within the social and natural sciences at the beginning of the twentieth century.

For Malinowski it was essential that anthropologists raised the scientific criteria on which they collected, analysed and represented their data. To this end, he identified ethnographic fieldwork as the surest method for studying human diversity. In a sense, the Trobriand Islands would become the laboratory where he would develop his research programme and transform the observational practices of all future anthropological projects. As his friend and fellow anthropologist James Frazer recounted in his preface to Malinowski's book, it was only by living among the natives, watching their daily routines and conversing with them in their dialect, that Malinowski was able to acquire a comprehensive understanding of the islanders.[3] Fieldwork was therefore identified at the beginning of the twentieth century as a necessary component of all anthropological research. This period in anthropology has since been canonized as the founding years of the modern discipline.

Ever since Malinowski visited the Trobriand Islands in the 1910s, engaging in ethnographic fieldwork has been a 'rite of passage' for anthropologists and a defining practice distinguishing them from other social scientists. Time spent on the spot allowed the new 'participant-observers' to claim an authoritative understanding of the culture under study and the ability to provide truly objective reports. However, as the discipline became 'fieldwork-centred', other traditional forms of observational practices – particularly those developed in the nineteenth century – became routinely marginalized within the research field. At the same time, since fieldwork's introduction as the cornerstone of the discipline at the beginning of the twentieth century, practitioners have argued that for a study to be considered an anthropological inquiry, it must be based to some degree on field experience. Akhil Gupta and James Ferguson have argued that 'all graduate students in social/cultural anthropology know, it is fieldwork that makes one a "real anthropologist", and truly anthropological knowledge is widely understood to be "based" (as we say) on fieldwork'.[4] Others such as Michael Angrisino and Julia Crane have stated that 'Practicing anthropology may include many experiences, but fieldwork, in particular, is often considered a sort of "rite of passage" and a necessary prerequisite for one to be considered a bona fide anthropologist'.[5] James Urry noted in his 1984 article 'A History of Field Methods' that

> Fieldwork holds a position of crucial importance in modern anthropology. Through fieldwork the discipline is supplied with new ethnographic information, on the basis of this material theories are developed and ideas are tested; by 'doing' fieldwork new anthropologists undergo a *rite-de-passage*, living and participating in alien cultures and thereby are admitted into the discipline.[6]

Where did this fieldwork ideal come from, and why did it become anthropology's chief preoccupation in the early twentieth century?

In order to fully comprehend the complexities behind the shift towards anthropology as a field-based science, we should first of all consider the state of the discipline at the turn of the twentieth century. As Henrika Kuklick has argued, developments in anthropology at the end of the nineteenth century 'can best be understood if we situate its practitioners within a larger community of scientists' because it was during this period that 'anthropologists then embraced field methods as one element in their strategy of adaptive accommodation to the intellectual ecology of the day'.[7] As the end of the nineteenth century approached, the various sub-disciplines of natural history continued to redefine their boundaries and create distinct and specialized methodologies. Responding to these disciplinary changes, anthropologists sought to strengthen the credibility of their observational practices by privileging first-hand experience as the most reliable way to acquire empirical evidence. Anthropology gave 'the field' a central position within its discourse.[8] Since then, as Gupta and Ferguson have put it, 'the single most significant factor determining whether a piece of research will be accepted as (that magical word) "anthropological" is the extent to which it depends on experience "in the field"'.[9]

While many scholars have long since abandoned the notion that fieldwork is a defining aspect of all anthropological studies, the rhetoric for the centrality of the field within the discipline continues to persist. This has created several problems within the historiography. For one thing, it has become conventional to depict the history of anthropology as fragmented into divergent methodological epochs. As an example, Barbara Tedlock – who has written extensively on ethnographic field practice – has argued that 'The Mythic History of anthropology is populated by four archetypes: the amateur observer, the armchair anthropologist, the professional ethnographer, and the "gone native" fieldworker'.[10] Rather than challenging this mythologized depiction of a disjointed discipline, Tedlock endorses it, emphasizing a pronounced division between the contributions of armchair-based 'amateur observers' from the late eighteenth and early nineteenth century, and the field-based 'professional ethnographers' from the first half of the twentieth century. In her version, informants such as missionaries, colonial officers and explorers were the sole providers of material for the armchair cogitations of nineteenth-century practitioners. It was not until the 1910s that academically trained ethnographers travelled abroad to undertake a recognizable form of intensive fieldwork and collect their own data.[11]

This characterization of the history of anthropological practices in such a manner has created a discontinuity between the methodologies of researchers interested in human diversity from the first half of the nineteenth century and the techniques utilized by field-based anthropologists from the early twentieth century onwards. In particular, scholars interested in the development of anthropological methods have positioned field techniques as an innovation of the early twentieth century, and Malinowski has been identified as a key instigator behind this disciplinary shift. Notably, James Urry has argued that 'this emphasis on Malinowski has helped to obscure the gradual development of field methods in the era before 1922, when Malinowski issued his famous challenge for scientific fieldwork'.[12]

This book goes further to argue that the overwhelming emphasis on the field has obscured a central component of all anthropological research, namely observation. The history of British anthropological practices is one of gradual change and of the adoption of new observational techniques into its methodologies. In consequence, this book builds on recent scholarship in challenging the traditional view of anthropology as a strictly field-based science by examining the history of observational practices within the discipline during the first half of the nineteenth century. Early ethnological and anthropological practitioners were acutely aware of the limitations of their methodologies and continually sought to refine their observational frameworks. Thus I will argue that fieldwork was intended to enhance the quality of anthropologists' observations and should be conceived of as an extension of earlier disciplinary practices.

Given the focus of this book, there are two kinds of scholars who will be reading the material in the subsequent pages: historians and social scientists. This is understandable given that the focus of this book is on the disciplinary history of anthropology. There are four major themes that connect the different chapters, and aspects of each of them will appeal to both readerships. In the conclusion of this book I discuss at length the details of how my four major themes connect to the five cases studies and bring them together. I will also show how this examination relates to other scholarly works within the history of science and anthropology. Nevertheless, it is important to consider each of them here briefly. In doing so, it will help to better situate this work with larger secondary literature.

The first theme is training, and the kinds of instructional preparations a researcher undertakes – regardless of their discipline – has a significant influence on their research practices. It shapes the way practitioners collect, analyse and represent their materials. In what follows I will examine the training regimes of nineteenth-century researchers interested in human diversity and link these activities to the larger picture of nineteenth-century science and medicine. The second theme is disciplinary development, and this has been a major area of research within the history of science and anthropology. As we will see in due course, the way nineteenth-century researchers went about establishing and transforming their research practices was similar to the activities occurring within other fields. The third theme within this work has to do with the importance of location and how a researcher's locality has a direct impact on their theories and methods. There are different benefits and limitations to living in London in contrast to working as a colonial agent in the South Pacific or Africa. Recognizing how these surroundings shape a researcher's observational practice adds greatly to our understanding of nineteenth-century human variation studies. Finally, the fourth major theme within this book is vision. The way nineteenth-century ethnologists and anthropologists went about observing human variation in a specialized way is an essential part of their techniques. How one trains the eye to see the world for specific purposes has always been a preoccupation for researchers within all scientific disciplines. With these four themes brought to the fore, it is my intention to reconsider the making of British anthropology and its relation to the context of Victorian science. There are many parallels that can be drawn between the disciplinary transformations occurring with anthropology during the nineteenth century and those happening within other research fields.

The remainder of this introduction will focus on two tasks. Firstly, it seeks to uncover the reasons behind twentieth-century anthropologists' discrediting the work of their Victorian counterparts. As we shall see, framing nineteenth-century researchers as 'amateurs' and 'armchair theorists' was part of the vocational and disciplinary strategies of twentieth-century anthropologists who were trying to secure their positions within the academy. The second objective of this introduction is to bring together anthropology's different methodological epochs by showing how observational practices have always been a chief preoccupation of researchers. In both sections, a principal theme is disciplinary development. How anthropologists have envisaged their science, and how they have practised it, is centrally important for this study.

## The Origins of Armchair Anthropology

In Kuklick's article 'Personal Equations', she recounts for her readers how anthropology shifted towards prioritizing field-based research as the surest method of observation at the beginning of the twentieth century. She identifies several changes that occurred during this transitional period which still shape our perception of the history of anthropological science today.[13] Building on her examination, I want to further consider how the vocational strategies of twentieth-century practitioners have misrepresented Victorian anthropology as primarily an armchair pursuit. There are two important episodes in the history of anthropology that have contributed to the misleading characterization of early observational practices within the sciences of Man – as it was known during the period – and the devaluation and rejection of Victorian anthropological research.

In the first instance, during the 1910s and 1920s changes in funding afforded many disciplines an opportunity to develop new research techniques. Prior to the institutionalization of disciplines such as anthropology at the start of the twentieth century, it was exceedingly difficult to procure funds for extended research trips abroad. However, with more money invested in academia, there was financial support available for, among other things, prolonged ethnographic research.[14] Anthropologists such as Malinowski and Baldwin Spencer (1860–1929) were some of the earliest to benefit from this economic and institutional upturn. At the same time, Malinowski and others of his generation sought to discredit Victorian anthropological practices as outdated and amateurish. Kuklick explained that 'as the natural history specialties differentiated, their practitioners determined that naturalists must break their long-established habit of relying on theories articulated by armchair scholars, that scientists could not do credible analysis unless they had themselves gathered the data on which their generalizations rested'.[15]

Malinowski envisaged a professionalized discipline of anthropology that worked cooperatively with Britain's imperial administration, and in consequence, he wanted to weed out untrained informants living in the colonies in favour of university-educated anthropologists. To this end, it was important to emphasize that anthropologists possessed important and unique skills that could help colonial officials develop imperial settlements.[16] According to Malinowski, only expert anthropological observers with university training would be able to accurately record and interpret the customs and values of indigenous peoples. In place of untrained informants collecting data *in situ*, anthropologists, with their specialized knowledge, conducted more rigorous and sophisticated ethnographic field studies. The primary aim was to acquire a more reflexive understanding of the culture under investigation, with special attention to local meaning systems. Frederick Erikson has outlined what early twentieth-century anthropologists believed to be the main benefits of having a trained ethnographer conduct field research:

> The ethnographer combined firsthand experience with an awareness of other forms of social life beside his own. What resulted, at best, was (1) more accurate descriptions of all the essential partial aspects of a society, described with reference to the society as a whole and, at least implicitly, to other societies as wholes; (2) more systematic definition of the social whole and its parts in terms established by the then growing disciplines of sociology and ethnology; and (3) less ethnocentric explanations of 'strange' customs in terms of their intelligible functions and meanings in the society being described.[17]

In short, the better-funded anthropologists of the early twentieth century argued that their field-centred approach to observing indigenous populations and collecting data was more theoretically informed and accurate in its representations.

It was during this period of so-called professionalization that new field practices came to prominence. In the 1920s Malinowski developed one of his most important practical contributions to anthropology: participant observation. Tedlock noted that 'Ever since Malinowski ... suggested that an ethnographer's goal should be "to grasp the native's point of view, his relation to life, to realize his vision of his world," there has been an expectation that participant observation would lead to "human understanding" through a fieldworker's learning to think, see, feel, and sometimes even behave as a native'.[18] This introduction of participant observation as the cornerstone of early twentieth-century anthropology helped to cement the privileged position of the field for anthropological research. As Anna Grimshaw has recently noted, 'The revolution which Malinowski claimed as his own established new goals for his followers. They set their sights on a position as scientists within the academy; and, in their drive for professional recognition, these new scholars sought to effect a radical break between past and present'.[19]

The second major rhetorical shift in anthropology, which further discredited the discipline's past, took place during the post-colonial period of the 1960s and 1970s. By this time the partnerships with colonial governments that Malinowski and others had promoted as ideal for the advancement of the science had become widespread and problematic. Among the consequences was a new critical awareness of the discipline's history that recognized the damaging effect ethnographic research had on the implementation policies of colonial governments.[20] Anthropologists had benefited from their relationship with colonial administrators and often exploited government power to gain access to indigenous communities. In return, anthropologists had become the 'handmaidens of the empire', providing colonial officials with detailed descriptions of indigenous populations. In many cases this had a devastating impact on extra-European peoples. Roy Ellen noted in his book that

> [The 1960s was] ... a period when social anthropology was on the defensive in the face of criticisms of its historical and ideological role as the hand-maiden of colonialism and neo-colonialism, and when anthropologists themselves were becoming increasingly aware of the shortcomings of their discipline and its methods.[21]

As anthropologists entered into a period of heightened reflexivity, they considered the damage the discipline had inflicted upon the indigenous populations they studied. With the history of their discipline marred by British imperialism, twentieth-century practitioners were increasingly unsympathetic towards their predecessors. They argued that nineteenth-century anthropology lacked direct engagement with extra-European peoples and as such routinely led to researchers objectifying and de-humanizing their subjects.[22] Grimshaw explained that 'Observation was identified as a dominant trope in modern anthropology, one which leads the fieldworker to adopt "a contemplative stance", an image suggesting detachment, indeed voyeurism, "the naturalist" watching the experiment'.[23]

In many cases, social scientists writing about anthropology's history have judged the value of Victorian contributors by comparing their methods and theories to contemporary standards. The anthropologist Frederick Barth is exemplary in his privileging of field-based anthropological investigations while downplaying the significance of nineteenth-century practitioners. He stated that 'The scholarly achievements of this early generation of ethnologists were insignificant'[24] and furthermore that their practice was 'based on written sources, not direct field observations'.[25] He further misrepresented these early studies by arguing that in the first few decades of the nineteenth century there was not much interest in a scientific study of human variation in Britain: 'While there was a large public market for serious travel literature, travel authors looked to history and geography for their wider perspective, and the lives of "savages" did not receive much serious attention'.[26]

Historians have also argued anachronistically about early nineteenth-century anthropology. Some have claimed that until the introduction of field-based research, a recognizable form of anthropology did not come into fruition until the late nineteenth century, with developments such as Alfred Cort Haddon's anthropological activities at Cambridge in 1880s and the Torres Straits Expedition in

1898. According to Kuklick, it was not until 1884 that 'the field [of anthropology was] accorded recognition indicative of its acceptance as a pursuit for genuine scholars and men of position'.[27] In keeping with this view, Kuklick has also promulgated a discontinuous depiction of anthropological practices by fragmenting its history into different epochs. For her, the study of human varieties at the start of the nineteenth century was an armchair pursuit more akin to natural history, in sharp contrast to its twentieth-century field-based incarnation.[28]

For Kuklick, the 'quintessential armchair scholar' from the nineteenth century did not engage in a practical study of humans per se, because he relied on the observations of untrained informants. Fieldwork was seen as dangerous, dirty and unfit for gentlemen. Thus it was informants from the lower classes who collected data *in situ* and provided 'gentleman-naturalists' with material for natural-philosophical theorizing. She wrote,

> the theoretical aspect of scientific work was for the mass of gentlemen-naturalists a comfortable task, performed within the familiar confines of their studies. Whether elite scholars were concerned to classify and explain flower, insect, or human variation, they confidently based their generalizations on data gathered by a congeries of collectors.[29]

Kuklick's interpretation of early nineteenth-century studies on human variation is suggestive, but it requires further unpacking. Although the majority of early ethnologists and anthropologists were men of some degree of standing, the dangers of travelling abroad and getting one's hands dirty was not necessarily a central concern. Several early practitioners within the sciences of Man, including Thomas Huxley (1825–95) and Richard King (1811–76), worked as surgeons on board military vessels and experienced the sorts of conditions Kuklick argued were unfit for gentleman-naturalists. Charles Darwin (1809–82) was also an important contributor to anthropology, and he travelled abroad during the 1830s, serving as both the naturalist and gentleman companion to Captain Robert FitzRoy (1805–65) on the *Beagle* voyage.

Kuklick's categorization of gentlemen-naturalists as a term to describe a group of researchers interested in natural history (broadly construed) also requires further reflection. The spectrum of nineteenth-century practitioners that could be demarcated as gentlemen-naturalists could include independently wealthy figures such as Darwin, or down-and-out figures such as Robert Knox (1791–1862), who after his involvement in the West Port murders (1827–8) lost much of his social and professional standing within Britain and barely lived above a level of pauperism. As we shall see in Chapter 3, despite his loss of reputation from the late 1820s onwards, Knox was still considered to be an important – albeit controversial – contributor to medical and natural history discussions right up until his death in 1862. Kuklick has also portrayed figures such as Edward Burnett Tylor (1832–1917) and James Frazer (1854–1941) as passive observers of ethnographic material, at the mercy of colonial informants and disengaged from their collecting practices. However, this was not the case: both of these figures were highly attuned to the problems associated with using evidence collected by informants, and each actively organized and monitored ethnographic exchange networks throughout the empire.[30]

It is also significant to recognize that the importance of the field as we understand it today had yet to be established in the first half of the nineteenth century. Practitioners were more concerned with refining their methods and theories for analysing ethnographic data (generally) than with the amount of direct experience they had with indigenous peoples. In other words, how one made sense of the data was the key preoccupation. As Robert Kohler argued, for a naturalist such as George Cuvier, 'fieldwork had the advantage of direct and vivid impressions, but [Cuvier would] assert that for breadth of comparison and objective analysis the closet naturalist had the advantage'.[31] Within the confines of their study, these naturalists stockpiled evidence and conducted comprehensive cross-comparative analyses of materials. They would identify patterns within their data sets and discard information that looked distrustful. Following on from this point, many social scientists and historians have underemphasized the fact that most early practitioners interested in human diversity conducted ethnological and anthropological research in their spare time. In the first half of the nineteenth century it was virtually impossible to support oneself as a full-time researcher of human varieties, and so practitioners had other careers such as physicians or surgeons. Consequently, in many cases they were unable to conduct their own research abroad because of commitments in Britain. Using informants in the colonies to collect data was one of the most effective ways to compensate for this constraint.

Not all of the scholars who have examined early anthropological practices have focused on the apparent lack of field-based experience among Victorian researchers. James Urry has examined the history of the discipline by showing the gradual changes occurring with anthropological practices from the late eighteenth century to the present day. He demonstrated that the emergence of the discipline did not occur suddenly and was the product of a collective effort to transform the science over several generations. By thinking about anthropology's past in such a manner, Urry's aim was to understand diachronically different anthropological theories and methods within their historical contexts. He stated,

> The development of fieldwork, and of the techniques associated with such research, was a gradual process. It went hand in hand with other developments both within and outside of anthropology: a more critical approach to ethnographic sources, changing theoretical interests, the increasing professionalization of academic anthropology and the easy access to remote areas of the world due to the expansion of European colonial control and improved methods of communication.[32]

Building on Urry's analysis, this book will further challenge anachronistic depictions of Victorian anthropology, firstly by considering the whole observational activity practised during the so-called armchair period, and secondly by situating these observational practices within their historical contexts. It will demonstrate that British researchers interested in human diversity in the first half of the nineteenth century were actively engaged in refining and transforming the way they observed, analysed and represented human varieties.[33]

The reflexive turn in anthropology during the 1960s and 1970s had other consequences: once scholars began to consider the ways in which their practices affected the lives of indigenous peoples, the next step was to reassess how they represented them. By questioning the inherent biases that influence every anthropologist's interpretation of different cultures, researchers acknowledged the politics of writing embedded within any anthropological discourse.[34] This had an important effect on historians who wrote about the discipline's past. They became acutely aware of their own ethnocentrism and how they judged the practices of Victorian anthropologists by comparing nineteenth-century research to contemporary standards. Thus there has been another change within the historiography since the late 1960s, where historians of anthropology have begun to be more introspective about the history of anthropological methods.

George Stocking was one of the earliest historians to challenge anachronistic characterizations of anthropology's history, and he demonstrated in his polemical article, 'On the Limits of "Presentism" and "Historicism" in the Historiography of the Behavioural Sciences', two different interpretive modes for studying past social scientific practices. The first mode was 'presentism', which sought to 'understand the past for the sake of the present'. Researchers looked for methods and theories from earlier conceptions of a discipline to see if they held any value for contemporary scholarship. The second mode was 'historicism', which sought to 'understand the past for the sake of the past'.[35] Stocking argued that social scientists have a propensity to dislike the historicist approach because 'he tends to demand of the past something more: that it be related to and even useful for furthering his professional activities in the ongoing present'.[36]

In Stocking's most famous book *Victorian Anthropology* (1987), he returned to his discussion on presentism versus historicism and argued for a more reflexive approach to the study of anthropology's history that combined both approaches. Stocking wanted scholars to identify the concerns of contemporary social scientists while utilizing the methods of historians. He wrote,

> the very decision to study [Victorian] ... thinking in the first place must reflect our own present interests, whether these be 'historicist' or 'presentist' in motivation. For in addition to providing answers to questions they were asking, their thinking may provide answers to questions that we ask, whether these have to do primarily with what it was that they were thinking, or what it is that we think today about similar issues – or, perhaps, why it was that they were preoccupied with certain questions and neglected to consider others that have interested anthropologists since.[37]

Recently, Kuklick has also engaged in the presentism versus historicism discussion, and she supported Stocking's framework for a combined approach to the study of anthropology's past. As with Stocking, she recognizes the value historical concerns have for contemporary anthropology, and vice versa.[38] Thus scholars who have examined the history of anthropology have become acutely aware of how contemporary and historical concerns are central for any study of the discipline's past and how both approaches can contribute to important new insights.

Another significant shift that has occurred in the secondary literature is how anthropologists have begun to challenge the unity of the discipline and its practices. Ellen observed in the introduction to his book *Ethnographic Research* that there has never been a unified method for conducting anthropological research, and that there have always been competing modes for collecting and interpreting ethnographic data. He wrote, 'Social anthropologists have sometimes had an image of their profession which presents them as highly atomized practitioners working with a shared (although not always specified) set of assumptions and procedures'.[39] Ellen, however, challenged this image of the discipline as a unified and progressivist enterprise. Under his critical gaze, anthropology required a more reflexive analysis of its methods, which emphasized the varying approaches to studying human diversity.[40]

Such a reflexive approach to the study of anthropology's practices has opened the door for other social scientists to reassess the discipline's most cherished assumptions. Gupta and Ferguson have reconsidered the significance of field practices in the introduction to their book *Anthropological Locations* (1997) and wrote, '"the field" itself, the place where the distinctive work of "fieldwork" may be done, that taken-for-granted space in which an "Other" culture or society lies waiting to be observed and written ... has been left to common sense, beyond and below the threshold of reflexivity'.[41] This was an important first step towards challenging the hegemony of the field within anthropological discourse because it promoted the idea that practitioners within the discipline should question its most taken-for-granted assumptions and reflexively consider the ways in which they have shaped the discipline's understanding of its practices and theories. Gupta and Ferguson went on to state, 'It is as if the mystique of fieldwork were too great in anthropology for the profession even to permit such obvious and practical issues to be seriously discussed, let alone to allow the idea of "the field" itself to be subjected to scrutiny and reflection'.[42]

However, Gupta and Ferguson struggled with their reflexive approach to studying anthropology's methodologies. Although they called for a reassessment of fieldwork as a set of practices and assumptions, they continued to conceive of it as the focal point of anthropology: 'The difference between anthropology and ... other disciplines ... lies less in topics studied (which after all, overlap substantially) than in the distinctive method anthropologists employ, namely fieldwork based on participant observation'.[43] I want to challenge this position and argue that the significance of fieldwork does not lie in its ability to connect its researchers directly with its object of study per se; rather, it is in its ability to further enhance the observational practices of anthropologists by providing them with another tool for collecting, analysing and representing different cultures. Now that I have outlined some of the central problems embedded in the historiography, I will turn my attention to the central aims of this book and discuss the significance of observational practices for the history of anthropology.

## Observational Practices and Anthropology

Simon Schaffer argued in his essay *From Physics to Anthropology and Back Again* (1994) that the history of anthropology is full of myths. Canonizing early twentieth-century figures such as Haddon, Malinowski and Franz Boas (1858–1942) as revolutionists within the discipline has obscured the story of anthropology.[44] The disciplinary transformation, which scholars have associated with the early twentieth century, is misleading because it blurs the connection between nineteenth-century and twentieth-century methodologies. Moreover, positioning Victorian anthropology as an armchair pursuit, without critically engaging with the meaning of this category, is particularly problematic because it has led to historical accounts that have further divided the history of the discipline into divergent methodological epochs.[45]

Schaffer, however, shifts away from this historiographical tradition by showing how figures such as Haddon were not ground-breaking researchers for the discipline, but rather they built their observational techniques on pre-existing methodologies. Thus Haddon's training in zoology both inside the laboratory and outside in the field played a critical role in shaping his observational practices.[46] This book builds upon and moves beyond Schaffer's analysis by tracing the transformations of observational practices within the sciences of Man during the first half of the nineteenth century. It further demonstrates the continuities that exist between Victorian studies on human diversity and twentieth-century anthropology. At the same time, it argues that there is still much to learn about the history of anthropology when one begins to look at the whole activity conducted during the so-called armchair period of the discipline.

Kuklick and others have suggested that the study of humans in the first half of the nineteenth century was a type of natural history. However, such a characterization requires some unpacking. Studying natural history was not a passive observational practice, where naturalists simply anthologized the materials collected by others. Naturalists were highly attuned to the problems associated with using second-hand reports. To ensure the quality of their data, naturalists built an active relationship with the various informants who were gathering evidence abroad. Practitioners of the sciences of Man were no exception; they were acutely aware of the sorts of information they required for their studies on human varieties and responded by directing the collecting methods of their informants.[47] Moreover, there are many instances where these researchers collected their own ethnographic data and engaged directly with their object of study. Darwin, Huxley and Tylor all travelled abroad and observed first-hand extra-European peoples *in situ*. If we are to expand our understanding of natural history, we need to examine in detail the observational practices of naturalists from each subfield. As with botanists and zoologists, practitioners of the sciences of Man possessed highly specialized observational techniques for analysing materials relating to human diversity. These materials not only included anatomical and physiological evidence, but cultural phenomena such as religious artefacts and detailed documentations of different languages. The disciplinary division between physical anthropology and its socio-cultural counterpart was not clear-cut in the nineteenth-century, and many ethnological and anthropological accounts included aspects of both subfields.

Building on the work of historians such as Lorraine Daston, Peter Galison, Daniela Bleichmar and Anne Secord, the main focus of this book will be to examine the observational practices of nineteenth-century researchers within the British sciences of Man. It outlines a more diachronic understanding of the history of anthropological practices. Rather than thinking about the discipline in terms of having two main diverging epochs, where Victorian anthropology was an armchair pursuit and twentieth-century anthropology was a field-based science, we should consider what is continuous within the discipline, and observational practices fulfil this requirement. Field-based research aimed to enhance the quality of anthropologists' observations, and as such it should be conceived of as a continuation and expansion of earlier disciplinary practices. In addition, some degree of armchair observation is still present within the field-centred discipline today. Before embarking on any fieldwork trip, it is customary for anthropologists to read the work of other scholars who have travelled to the location.

What it means to observe something does not simply entail the physical act of looking at something. Rather, it is a much more specialized practice of analysing and interpreting an object or specimen's meaning and significance. As Bleichmar recently noted, the observational practices of naturalists did not simply involve 'sight, but rather insight'.[48] All forms of specialized observational practices include laborious training, and practitioners from any discipline develop discriminating practices that seek to identify those characteristics an object of study possesses which are of importance to researchers.[49] As Daston and Galison argued in *Objectivity* (2002), 'All sciences must deal with the problem of selecting and constituting "working objects ..."'.[50] By selecting rudimentary materials on which to base foundational training, newcomers to any discipline learn to observe the world in a specialized way. In the case of ethnology and anthropology, Urry has noted, 'ethnography does not attempt to "record" the totality of everyday life in a particular context, but is the result of a process of selection by the anthropologist according to the needs of interpretation, explanation, and generalization'.[51]

The type of professional training and institutional backing researchers receive is a major influence on their observational method. As an example, many ethnologists and anthropologists during the first half of the nineteenth century studied medicine, and this background fed into their writings on human diversity. Anatomical and physiological topics were a staple of ethnological literature. Medically trained figures such as James Cowles Prichard (1786–1848) and William Lawrence (1783–1867) focused on anatomical and physiological features – especially skull conformations – to distinguish the various races.[52] The medical world also influenced ethnological observational practices in other ways. Anatomical lecture theatres in cities such as London and Edinburgh provided early ethnologists such as Lawrence and Knox with a venue for disseminating the edifying principles of the emerging science to a new generation of practitioners.

Colonialism was equally significant for facilitating ethnological and anthropological observation. The British Empire, with its settlements scattered across the world, possessed a global network for gathering and exchanging information. This allowed ethnologists and anthropologists in the metropole who were unable to travel abroad to acquire data on many different races. Huxley was one of the earliest practitioners to take advantage of this infrastructure, and he organized with the Colonial Office in London a scheme that encouraged officers living abroad to photograph indigenous people for the benefit of the sciences of Man and the empire. In turn, observational practices were influenced by this partnership with the British government because ethnological and anthropological research had to show its utility for colonial administrators. There are also other connections between anthropology and imperialism, and many practitioners during the first half of the nineteenth century took up an interest in human diversity because they encountered extra-European peoples abroad while working as colonial agents.

In all disciplines, experienced researchers produce instructive literature that explains the key principles and methodologies of the field of study. Whether these texts are on botany, geology or ethnology, they all share the same purpose, which is to develop more refined observational skills. Daston and Galison, for instance, focused on scientific atlases from the eighteenth and nineteenth century and remarked that

> Whether atlases display crystals or cloud chamber traces, brain slices or galaxies, they still aim to 'map' the territory of the sciences they serve. They are the guides all practitioners consult time and time again to find out what is worth looking at, how it looks, and perhaps most important of all, how it should be looked at.[53]

Observational practices do not stop there, and they embody many different modes of visual refinement. Skilled practitioners from any discipline learn to describe, analyse and represent the natural world according to the guiding principles of the research field's methodologies. Thus not only do practitioners learn to perceive their object of study in a specialized way, they also refine their analytical and writing practices. Different disciplines have specific terminology for describing characteristics or explaining phenomena. In the case of ethnology and anthropology, natural history taxonomical vocabularies and systems were important parts of a researcher's observational gamut. By organizing and describing different indigenous populations into groupings based on physical and cultural characteristics, early practitioners within the sciences of Man believed they could explicate theories regarding the origin of humans.

As still holds true today, every research field had specialized training grounds, resources and instruments in the nineteenth century. Early ethnologists and anthropologists learned about their object of study and further developed their observational techniques in locations such as lecture theatres, medical and natural history programmes at universities, museums and learned societies. The instruments these researchers used in their studies not only included devices, such as measuring equipment for recording weight and height, but also assistants, who functioned as types of tools.[54] In the case of the nineteenth century, colonial-informants collected data for ethnologists and anthropologists living in the metropole, and these workers were an extension of the researchers' observational arsenal.

Another important component of observational practice has to do with the issue of trust. Practitioners within any research field consider certain sources of knowledge more credible than others. In particular, scholars since the seventeenth century have prioritized knowledge based on first-hand experience as the most reliable source of scientific knowledge. Steven Shapin argued in his book *A Social History of Truth* (1994) that 'seventeenth-century and present-day "moderns" widely advertise direct experience as the surest grounds for factual knowledge, just as they identify reliance upon the testimony of others as an insecure warrant for such knowledge'.[55] Shapin wanted to challenge this conception of science, and he argued that most scientific knowledge is based to some degree on the observations and research of other practitioners. As a result, all forms of knowledge are collective endeavours. He wrote, 'Knowledge is a collective good. In securing our knowledge we rely upon others, and we cannot dispense with that reliance'.[56] Building upon Shapin's reconsideration of second-hand sources within knowledge discourses, I argue that a reliance on the testimony of other researchers does not necessarily devalue the significance or credibility of a work.[57] Thus observational practices encompass a wide range of methods, and as Secord argued in her thesis *Artisan Naturalist* (2002), 'speaking, reading, writing, counting, looking, [and] walking' were all part of the naturalist's framework.[58]

For my purposes, I am going to approach the study of nineteenth-century anthropological observational practices by combining presentist and historicist perspectives. Secord suggests that 'Sociological and anthropological models can help in the reconstruction of past practices' because they provide scholars with new questions for studying history.[59] I will build upon Stocking's and Kuklick's reflexive models by challenging ethnocentric historical examinations that devalue the contributions of Victorian researchers interested in human variation. I will do so by demonstrating what is continuous between the sciences of Man in nineteenth century and modern anthropology: a desire to transform and refine the way in which practitioners observed, analysed and represented their data. In doing so, I will seek to contextualize these observational practices by framing them as products of their historical periods.

With this in view, I will be conscious of how twentieth- and twenty-first-century scholars have misinterpreted anthropology's past, while at the same time attempting to make sense of older practices by representing them within a historicist framework. In some cases, this method of analysis will require me to break down the observational practices of nineteenth-century researchers by using contemporary language and terminology. This will enable me to demonstrate the common links between the older techniques of ethnologists and anthropologists from the nineteenth century, and modern practitioners from the 1910s onwards. In doing so, it will strengthen my argument for the continuity that existed within the so-called disciplinary epochs of the research field. Ultimately, the aim is to demonstrate that the focal point of anthropology has always been its observational practices and that the field is an extension of this preoccupation.

## Five Case Studies

In his 1992 essay 'Retrospective Prescriptive Reflections', Stocking discussed how he struggled in the earlier part of his career with writing 'big picture history', and how he often favoured an approach that examined different 'vignettes' which occurred during the history of the discipline because it allowed him to examine the 'multiple contextualizations' of the anthropology's past.[60] Such a method is particularly useful for a close study of anthropological theory and method. Similarly, this book uses five case studies to explore different themes connected to nineteenth-century observational practices within the British sciences of Man.

In Chapter 1 I consider the disciplinary origins of ethnology and look at the work of James Cowles Prichard and William Lawrence. By considering both figures' training in natural history and medicine, I will discuss how these disciplinary traditions influenced their vocational strategies and ethnological writings. Both Prichard and Lawrence built their methodologies on pre-existing observational frameworks such as Carl Linnaeus's classificatory techniques,

Johann Friedrich Blumenbach's craniometry and Captain James Cook's travel narratives. I will argue that books such as Prichard's *Researches into the Physical History of Man* (1813) and Lawrence's lectures on the *Natural History of Man* (1819) aimed to improve the observational practices of early practitioners of the sciences of Man. Finally, human origins were a chief preoccupation of ethnologists throughout the period under study, heavily influencing practitioners' observational techniques. Thus I will examine at length both Prichard's and Lawrence's monogenetic theories regarding the origin of races and show how it shaped their observational paradigms.

In Chapter 2 I look at the development of ethnographic manuals and questionnaires in the 1830s and 1840s and show how British ethnologists sought to improve the collecting practices of their informants working abroad. I will begin by looking at the development of specialized societies devoted to the scientific study of Man. I will also demonstrate how the Ethnological Society of London (f. 1843) based many of its guiding principles and initiatives on the early activities of French societies such as the Société des Observateurs de l'Homme (f. 1799) and the Société Ethnologique de Paris (f. 1830). Moreover, members of the Ethnological Society of London based the structural format and content of their questionnaires on the earlier French literature. I will also examine the efforts of the British ethnological community to establish a global network of researchers and informants who would work cooperatively in developing the scientific study of humans.

Chapter 3 considers the ethnological debates of the 1850s and discusses the work of Robert Gordon Latham (1812–88) and Robert Knox. I will be offering close readings in particular on Knox's treatise *Races of Man* (1850) and Latham's *Natural History of the Varieties of Man* (1850). Why these two figures? Undoubtedly they had much in common, most obviously their interest in the natural history of human varieties and their ancestral origin. What is more, both are familiar within the historiography, having been canonized as founding fathers of British anthropology by early practitioners, periodical reviewers and historians of anthropology. However, many social scientists and historians fail to appreciate that most early practitioners of the sciences of Man conducted ethnological and anthropological research in their spare time, and that these researchers had careers that tied them to Europe. To compensate for their inability to conduct their own research abroad, both Latham and Knox developed innovative observational techniques for engaging with their object of study in Europe. Knox conducted a type of field study locally by examining Gypsy and Jewish populations in Britain and in the Netherlands, while Latham collected philological evidence in Scandinavia and created an ethnographic training site at the Crystal Palace in London.

In Chapter 4 I reconsider the anthropological debates of the 1860s by looking at the observational frameworks of James Hunt (1833–69) and Thomas Huxley.

Most historiographic accounts have positioned the debate as a competition over practice, between anthropologists led by Hunt on the one hand and ethnologists led by Huxley on the other. However, I will show that there was more in common between the two sides than the rhetoric of the historiography suggests. I will demonstrate that Hunt and Huxley had similar views with regards to race and women, and each figure sought to improve the ways in which researchers collected and analysed their data. Hunt attempted to establish anthropological societies around the world where their members would collect ethnographic data and send it to the Anthropological Society of London (f. 1863). Conversely, Huxley worked collaboratively with the Colonial Office in London and provided instructions for colonial officers to use for photographing and cataloguing human diversity in their settlements.

Finally, in Chapter 5 I consider the work of Charles Darwin and Edward Burnett Tylor and show how both figures continued to transform and refine the discipline's observational practices by bringing together many of the research methods discussed in the previous chapters. I will examine in detail the observational paradigms discussed in Darwin's *Descent of Man* (1871) and Tylor's *Primitive Culture* (1871). In both cases I will demonstrate how their works from the 1870s were syntheses of the various techniques utilized throughout the first half of the nineteenth century, from questionnaires and informants to direct *in situ* observation abroad. In the case of Darwin I will begin by showing how his time on board the *Beagle* voyage influenced his writings on humans. Tylor also possessed first-hand experience travelling and living among indigenous peoples, and his time in Mexico was an important part of his development as an ethnologist. In the final section of this chapter, I will discuss both Darwin's and Tylor's instructions for informants.

It is an important reflexive exercise for all historians, regardless of their research interests, to think critically about why their work matters and how it fits into the larger historiographic picture. In my Conclusion I do just that: I consider how my focus on observational practices will help to redefine the disciplinary history of British anthropology, and I show how this story connects to the larger context of nineteenth-century science. Anthropologists have always been preoccupied with the ways in which they have observed, analysed and represented human varieties. By bringing these techniques to the fore, this book shows how practitioners throughout the nineteenth century continually sought to transform and refine their observational methods. This preoccupation was similar to what was occurring in the other natural sciences, such as botany, zoology and geology.

The disciplinary shift that occurred at the beginning of the twentieth century, which saw anthropology reposition itself as a field-based science, was part of a longer process that aimed to improve the way its researchers engaged with their object of study. In this book I look at the whole observational activity conducted during the first half of the nineteenth century, emphasizing its commonalities with later twentieth-century methods. The introduction of fieldwork as the cornerstone of the discipline in the 1910s onwards emerges as one more tool added to the arsenal of the skilled observer. In the Conclusion I also draw together some of the overarching themes from the book, including anthropology's relationship to medicine and imperialism, disciplinary development and professionalization.

There is one final question that I would like to examine before concluding this Introduction: what is the correct disciplinary label to use when discussing the study of human diversity in the first half of the nineteenth century? Because so much of this book has to do with disciplinary formation, it is important to think about the terms researchers used to describe their science. This is a topic that I will return to throughout the various chapters, but it is useful to begin considering it here briefly. Before the formation of the Anthropological Institute of Great Britain and Ireland in 1871, practitioners in Britain had yet to assign a single designation to the study of human variation. Early researchers used several different names, and this is indicative of a science still in its formative years. In view of this point, I will employ several different terms to refer to the discipline in an effort to demonstrate its transitional state in the first half of nineteenth century.

One of the earliest terms that practitioners used to describe the discipline was the 'Natural History of Man'. It appeared in the pioneering works of both Prichard and Lawrence in the 1810s, and it is suggestive because it shows that some early researchers saw the study of human varieties as a subfield of natural history.[61] Interestingly, the terms 'Anthropology' and 'Ethnology' – the two disciplinary labels most commonly ascribed to the study of human diversity – were not widely circulated in Britain until the 1850s and 1860s. The *Penny Cyclopaedia* from 1837 was one of the earliest texts to define the term 'Anthropological', and it had two meanings for the word: the first was '*Anthropological Didactic*, or instructions for learning both the Interior and Exterior of Man', while the second definition was '*Anthropological Characteristic*, or the way to find out the Interior from the Exterior [of Man]'.[62] Both of these explanations were ambiguous, indicating the vague conceptual notion of the research field during this period. Moreover, until the 1860s 'Anthropology' was not widely used as the designation of the discipline. Its usage became more common in Britain when James Hunt and his associates touted the term as the most accurate word for characterizing the physical study of races.[63]

'Ethnology', on the other hand, first appeared in the *Oxford English Dictionary* in 1842, and although its inclusion into the discipline's lexicon came later than other designations, it quickly became the favoured label for many practitioners.[64] Prichard has been credited as the first researcher in Britain to use the term, and it was mentioned in his book *The Natural History of Man* from 1843. Because Prichard attempted to link all the races of the world to a single origin based on common linguistic features in their lexicons, 'Ethnology' has been routinely associated with the non-physical study of human variation.[65] Finally, another term commonly ascribed to the discipline in the early nineteenth century was the 'Science(s) of Man'. It was positioned as a neutral designation and was often used by writers who did not want to align themselves with any specific methodological focus. Because there was no consensus as to what to call the discipline in the first half of the nineteenth century, I will use all of these designations in this book. With this in mind, when examining the methodologies of my historical actors, I will attempt to be as accurate as possible and employ their terminology. It is highly instructive that there was no agreement between nineteenth-century practitioners on what to call human variation studies. It shows that the discipline of anthropology was in its formative years and that key aspects of the research field were still being negotiated.

# 1 FOUNDING THE SCIENCES OF MAN: THE OBSERVATIONAL PRACTICES OF JAMES COWLES PRICHARD AND WILLIAM LAWRENCE

> The question [of human origin] belongs to the domain of natural history and physiology; we must be contented to proceed in our examination in the slow and humble, but sure method of observation.[1]
>
> William Lawrence, 1819

Observation has been central to studying human diversity from its earliest days as a field of inquiry. The ability to identify and organize different ethnicities into groupings based on similarities of form and structure was the cornerstone of ethnological inquiries. Since the eighteenth century, naturalists, medical practitioners and explorers had been developing increasingly sophisticated modes for observing and interpreting the natural world and its inhabitants. Early ethnologists such as James Cowles Prichard and William Lawrence built their analytical frameworks for studying humans on these well-established observational traditions. Taking their contributions as its focus, this chapter seeks to enlarge our understanding of early ethnological practices by looking at how researchers at the beginning of the nineteenth century based their observational techniques for studying human diversity on natural history, anatomy and physiology, and travel writing. Prichard and Lawrence are good examples for this examination because their research utilized aspects from all three traditions.

Prichard is a difficult case because he has traditionally been positioned as a passive observer who anthologized the reports of others. However, a close examination of his work shows that he was a skilled scientific observer who actively utilized a highly sophisticated methodology for collecting, analysing and representing his ethnographic material. Lawrence, on the other hand, has routinely been characterized as an important contributor to natural history and medicine in the first half of the nineteenth century. His anatomical and physiological observations are widely discussed in the secondary literature.[2] By comparing the methodologies and theories of both of these figures, we see many similarities between their observational frameworks. This further strengthens the argument that Prichard and Lawrence were highly attuned to the problems associated with studying human variation in the early part of the nineteenth century, and that the emergence of the discipline was rooted in concerns for scientific accuracy and credibility.

Prichard and Lawrence are also good examples for this study because they are familiar within the historiography. They have been canonized as founding fathers of British anthropology by early practitioners, periodical reviewers and historians. In 1848 the physician, biologist and critic for the *Edinburgh Review*, William Benjamin Carpenter (1813–85), expressed his 'high sense of the value [for] ... the labours of Dr. Prichard; who has unquestionably done more than any other single individual to place Ethnology on a scientific basis'.[3] Carpenter continued by stating that 'Dr. Prichard has acquitted himself in each [pursuit], whether Physical Geography, Anatomy, Physiology, Psychology, History, or Philology ... [and] we are sure that he may rank as *facilè princeps* among those who have attempted to bring them all into mutual relation'.[4] Reviewers also positioned Lawrence's work in the same light. In 1822 the *Imperial Magazine* wrote, 'The high reputation and general acknowledge ability of Mr. Lawrence, are sufficient, of themselves, to give a more than ordinary degree of the importance and interest to any subject which he should select as the object of his attention'.[5]

This chapter is divided into four main sections and traces the emergence of ethnological observational methods in the early decades of the nineteenth century. In the first section, I look at the careers of Prichard and Lawrence in turn and show how their training in natural history and medicine influenced their writings on humans. This will show how their ethnological paradigms integrated within their vocational strategies. This was significant because in the 1810s there was very little work written about human variation in Britain, and both Prichard and Lawrence were able to forge positions as founders of the emerging research field. In sections two to four, my focus will shift to a close examination of the first edition of Prichard's *Researches into the Physical History of Man* (1813) and the reprinted 1828 edition of Lawrence's lecture on the *Natural History of Man* (1819).[6] Through a close textual analysis, I will demonstrate how natural history, medicine and travel narratives shaped their ethnological writings.

## Learning to Observe: Prichard and Lawrence

> [Prichard's] patient industry in collecting materials, and the many branches of knowledge he has brought to bear upon his subject, have rendered him eminent among the writers on Ethnography.[7]
>
> *Athenaeum*, 1842

### *James Cowles Prichard: Physician and Ethnologist*

In the first half of the nineteenth century, Prichard was one of the foremost writers on the subject of human diversity in Britain. He systematically categorized the different races that inhabited the world and formulated a monogenetic theory for their common origin. Classifying human variation was a complex process, and it required comprehensive training in several different research fields. Looking at Prichard's educational background is instructive because it demonstrates how his training in natural history, medicine, language and philosophy informed his ethnological observational paradigm. Moreover, Prichard's diverse interests – reflected in his ethnological model – are representative of the state of ethnology in the early nineteenth century. As a field of research, ethnology was in its formative years – in fact it was called the 'Natural History of Man' until the 1830s – and it brought together data from various sources and areas of study.

Prichard was born on 11 February 1786 in Ross-on-Wye in Herefordshire. Throughout his life he was a devout Christian, and this shaped his vocational path. His father Thomas Prichard (1765–1843) was a committed member of the Society of Friends, and James and his siblings had a pious upbringing. The Quaker belief in the unity of races had a lasting effect on his racial views, and throughout his long career writing on ethnological topics, monogenism – a theory postulating the common origin of humans – was a central tenet of his framework.[8] Because of his family's religious affiliation, Prichard received private tutoring from his father's home in Bristol, focused on language studies, and he learned French, Latin and Greek. As Hannah Augstein has pointed out, 'If Prichard had a talent for languages, his education furthered the natural inclination'.[9] Language would become a primary component of Prichard's ethnological research throughout his career, and he argued that different human varieties could be linked to a single origin by tracing commonalities among languages.

However, it was through medicine that Prichard earned a living, and between 1802 and 1804 his father secured for him several apprenticeships with surgeons and apothecaries.[10] It was also in 1804 that Prichard decided he wanted to be a physician. As a Quaker he was limited as to where he could study and receive a degree. Prior to the educational reforms of late nineteenth century, both Cambridge and Oxford used religious tests that restricted non-Anglicans from either entering (in the case of Oxford) or graduating (in the case of Cambridge).[11] However, there was another option for studying medicine in Britain, and in 1805 Prichard enrolled in the medical programme at the University of Edinburgh. It was an important developmental period for him, and his training in natural history and medicine would heavily influence his ethnological paradigm. Prichard would go on to develop an analytical framework for organizing humans into groupings based on their physical and non-physical characteristics using natural history classification methods and comparative anatomical practices.

By the early nineteenth century, Edinburgh had established itself as a preferred destination for Quakers wanting to train in medicine because it operated without any religious testing.[12] It was also a cheaper option than either Cambridge or Oxford and offered a high-quality medical education with an impressive list of teachers, including the famous lecturer of anatomy Alexander Monro Secundus (1733–1817),[13] and an anatomical lecture theatre and a hospital. With such teachers and facilities, students gained a comprehensive understanding of both medical theory and clinical experience. In addition to Edinburgh being a more liberal institution than either Cambridge or Oxford, it also had many ties to continental scholarship. The university had several affiliated extramural schools, which circulated the ideas of Johann Friedrich Blumenbach (1752–1840), Carl Linnaeus (1707–78), Jean Baptiste de Lamarck (1744–1829), Georges Cuvier (1769–1832) and Étienne Geoffroy de Saint-Hilaire (1772–1844).[14] It is evident that Prichard's exposure to these naturalists had an effect on his ethnology, as all five featured in the footnotes of the various editions to his *Researches*. In particular, Blumenbach's taxonomy of races was a fundamental component of Prichard's ethnological classifications.[15]

For those students interested in further developing their practical abilities, these extramural institutions also provided many young researchers with additional opportunities to work directly with ethnographic objects such as human skulls. Although we do not know for certain whether Prichard attended courses at any of these schools outside of the university, he was nonetheless living in a community where he was surrounded by this intellectual culture. Thus Edinburgh furnished Prichard with a sound knowledge of natural history and anatomy, both from his lectures at the university and from discussions and readings of continental scholarship from the extramural schools and learned societies. Combining his interest in these topics, Prichard decided to specialize in the study of extra-European peoples, and between 1807 and 1808 he wrote his MD dissertation on the varieties of human races.[16] This material was later reworked into the first edition of his *Researches*.

Upon completion of his MD in 1808, Prichard entered Trinity College Cambridge, where he was a 'Gentleman Commoner' until 1809, studying theology and mathematics. However, because of the limitations imposed on dissenter groups – such as the Quakers – during the first half of the nineteenth century, Prichard decided to convert to Anglicanism so that he could better integrate into the Cambridge scholarly community.[17] His conversion made it possible for him to further his study at Trinity College Oxford in 1810, where he studied several languages, learning Arabic, Celtic, Hebrew and Sanskrit. Acquiring these language skills would prove to be valuable for Prichard because it enabled him to use ancient sources in his later ethnological works.

By the end of 1810 Prichard moved to Bristol, where he set up a private medical practice. In 1811 St Peter's Hospital in Bristol hired Prichard as a physician to work in their lunatic asylum. In 1816 the Bristol Infirmary elected him to the post of physician, which he held until 1843.[18] Prichard devoted much of his time to the study of mental illness. Having worked in the lunatic asylum from 1811, he went on to publish several works on insanity, such as his *Treatise on Insanity* (1833) and *Insanity in Relation to Jurisprudence* (1842).[19] Moreover, between 1826 and 1828 he was a medical visitor to the Gloucestershire madhouses, which influenced much of his writings on the mind. In 1845 Lord Shaftesbury appointed Prichard to the post of commissioner for the Lunacy Commission in London.

Prichard's various positions within the medical community had an effect on his ethnological methodology. After all, it was virtually impossible to earn a living as an ethnologist in the first half of the nineteenth century, and he turned to a career in medicine to support his family. Part of the reason why Prichard was unable to travel abroad and collect his own first-hand observations of indigenous populations had to do with the fact that he was tied to his various medical posts in England. Thus he became reliant on the observations of travellers for his ethnological inquiries. It is partially for this reason that he has been positioned within the historiography as a passive observer who anthologized the works of other writers. That said, however, Prichard's observational practice was a highly sophisticated methodology that laid the foundation for several ethnological and anthropological approaches to the study of human diversity in the decades leading up to the Torres Straits Expedition in 1898.

## *William Lawrence: Physician, Lecturer and Ethnologist*

Historians of anthropology have for the most part undervalued William Lawrence's contributions to the study of human variation in the first few decades of the nineteenth century and have positioned him as a peripheral figure. However, his lectures on the *Natural History of Man* constituted one of the first scientific publications in Britain to specifically discuss human variation. Accordingly, he was canonized by early nineteenth-century practitioners as one of the founding fathers of the discipline.[20] As was the case with Prichard, Lawrence also had a background in natural history and medicine, and his training in both influenced his ethnological writings.

Lawrence's entry into medicine began in 1799, when he apprenticed as a surgeon under the tutelage of the lecturer of anatomy John Abernethy (1764–1831) at St Bartholomew's Hospital in London. Abernethy was regarded as one of the leading anatomists in Britain during this period, and under his instruction, Lawrence specialized in anatomical and physiological observation.[21] As was the case with Prichard, Lawrence also took an interest in continental scholarship. In 1807 he translated Blumenbach's *A Short System of Comparative Anatomy* from German.[22] This connection to Blumenbach is further significant because Lawrence based his classificatory system for ordering humans on the German physician's framework.[23] In addition, Lawrence dedicated his lectures on the *Natural History of Man* to Blumenbach, and he wrote,

> I think [it] is a mere act of justice to dedicate this Work to you, I do so with the greater pleasure, because it affords me the opportunity of gratefully acknowledging the instruction and entertainment which I have derived from your excellent writings; of recommending to imitation the example you have set, of combining together anatomical, physiological, and zoological pursuits ...[24]

Blumenbach's work was important for Lawrence as it outlined a method of observation that combined natural history and medicine. He would go on to base his ethnological framework on this model, incorporating the taxonomic practices of naturalists and the identification methods of anatomists.

It was virtually impossible to forge a career as a paid ethnologist in the early part of the nineteenth century, and seeking a viable vocational path, Lawrence earned a living in medicine. By 1815 his reputation as a leading British anatomist blossomed, and he was appointed Professor of Anatomy and Surgery at the Royal College of Surgeons. Lawrence was a charismatic speaker, and this afforded him with an opportunity to establish himself as a key disseminator of scientific and medical knowledge in London. However, his standing within the community began to wane soon after, when in 1816 he challenged the view of the eighteenth-century surgeon and anatomist John Hunter (1728–93), who argued that the human body contained 'an active power or principle, distinct from matter, which controlled the life processes, resisted bodily decay, and directed embryonic development'.[25]

As a result of challenging Hunter's view, many of Lawrence's contemporaries, including Abernethy, accused him of materialism.[26] Lawrence's most significant contribution to the study of human variation was his lectures on the *Natural History of Man*, which occurred at the start of his medical career. After he was branded a materialist, he produced no further major ethnological work. His lectures were withdrawn and banned, and only pirated copies were available. That said, Lawrence maintained a prominent position within the British medical world and even became the sergeant-surgeon to the queen by the end of his career in 1857. Lawrence's writings on the natural history of humans continued to influence ethnological and anthropological research long after his controversial critique of Hunter during the 1810s, and we can see traces of his ethnological methodology in Thomas Huxley's and Charles Darwin's work.[27]

Both Prichard's and Lawrence's training in natural history and medicine shaped the ways in which they observed and analysed topics relating to human diversity. Natural history would provide them with a system for organizing their material into manageable groupings for analysis, while anatomy and physiology taught them how to identify important characteristics in different races. Furthermore, in each case we can see how continental scholarship had an impact on Prichard's and Lawrence's work. Writing on human variation in the 1810s also afforded both figures an opportunity to forge positions as founders of a nascent research field. After all, prior to the publication of Prichard's *Researches* in 1813, and Lawrence's lectures in 1815–16 and its publication in 1819, there was very little literature strictly devoted to the natural history of humans in Britain.

## Ethnology and Natural History

> The natural character is made out by the most accurate description of the structification [*sic*] of the first species; then all the other species of that *genus* should be compared with the description of the first ...[28]
>
> Carl Linnaeus, 1775

### *Prichard, Natural History Classifications and Monogenism*

From the opening pages of the *Researches into the Physical History of Man*, observation was at the forefront of Prichard's methodological approach to the study of human diversity. In his introduction Prichard wrote, 'If an illiterate person, bred in some remote corner of England, who had never seen or heard of any human creatures different from the natives of his own vicinity, were suddenly transported into the western continent, and introduced to a horde of the naked and dusky barbarians who wander on the shores of the Mississippi ... he would certainly experience strong emotions of wonder and surprise'.[29] Using the example of a fictional British person, Prichard transported his figure to the colonies, where he was confronted with indigenous people who appeared physically different to the typical Europeans he was accustomed to seeing.

The point of this thought experiment was significant because Prichard was trying to get his readers to imagine themselves in similar situations and consider how they would interpret cross-cultural encounters with indigenous peoples. Prichard wrote that upon meeting these inhabitants, his British traveller 'would indeed immediately recognize the beings whom he saw as men, for the expression of rational intellect; the likeness of the Creator which was imprinted on the first of the human kind, is every where instantly striking and conspicuous'.[30] The next step for his 'fictional observer' would be to compare himself to the native population and interpret their differences. Prichard argued that his British traveller would conclude that the indigenous people were fellow humans of common ancestral origin, regardless of their differing physical appearance. He wrote, 'a spectator in such circumstances would be exceedingly perplexed in contemplating appearances so new to him, and in comparing with himself persons who differed from him in so extraordinary a manner, and yet so nearly resembled him, that he could not fail to consider them as fellow creatures'.[31]

The primary aim of Prichard's work was to formulate a system for understanding human difference, and the crux of this analytical process was a desire to demonstrate the unity of races. To accomplish this task, Prichard outlined two main components of his observational method. The first was to organize the various races of the world into a taxonomical system based on observable physical and cultural characteristics. He combined this classificatory practice with his anatomical knowledge. The second component was to analyse this data and demonstrate how the differences observed in these racial groupings were of 'variety' and not of 'species'. Prichard believed that all humans shared one ancestral origin, and he wrote,

> In the following pages I shall endeavour to state as clearly as possible the principal facts whether physiological or historical ... whether it is probable that all mankind are the offspring of one family, and shall afterwards proceed to trace the affinities of different nations, as far as an inquiry of this nature may tend to throw any light on the physical history of man. I shall in the course of this investigation endeavour to obtain some idea of the efficacy of those causes, which have been supposed capable of producing the divestitures of the human kind [*sic*][32]

This was the task that Prichard sought to achieve in his *Researches*. It was one of the earliest attempts by a British practitioner to systematically trace the differences between the races and articulate a theory that linked all of humanity to a single ancestry.

The issue of human origin was a complicated and heavily contested one in the early nineteenth century. As is well known, there were two main opposing viewpoints. Firstly, there were those later dubbed monogenists, who at a rudimentary level argued that all humans shared a common origin. Conversely, there were those later named polygenists, who generally speaking argued that the different races of the world came from separate origins. Prichard discussed both sides of the debate and stated that

> Some authors who have treated of this subject have supposed that all the nations on the earth are descended from a single family, and have attributed the varieties which we

> observe in their aspect and bodily structure to the action of natural causes on a race originally uniform. Others on the contrary insist that the differences adverted to, are too great to have been so produced, and must therefore be referred to original diversity.[33]

To resolve the problem, Prichard turned to natural history classification and argued that the most empirical way to determine whether all humans shared one ancestral origin was by establishing whether the different races of the world were varieties of the same species. He wrote,

> Our first and most important attempt of this pursuit must be the solution of the inquiry whether all mankind are to be considered as of one natural species, or not; or whether the physical diversities which so curiously distinguish the several races of men are, to borrow a term which is chiefly used in abstract reasonings, specific differences or only varieties.[34]

The main problem for Prichard was making sense of racial variation, and a key component of this analytical process was determining whether these differences seen in humans were so great that they should be separated into distinct groups.

Observation was central for this inquiry, and Prichard wanted to formulate a framework to facilitate the ordering of characteristics into clearly defined groupings based on similarities of structures. This enabled him to divide copious amounts of data into more manageable quantities for analysis. Prichard stated that

> The design of arrangements of this kind is chiefly to facilitate memory by a lucid order; to enable it to retain in its comprehension a vast variety of phænomena, in which in the confused mass it could have no secure hold, by distributing them according to certain classes. Such classes are constituted by some one character possessed by all the individuals belonging to each, or by a general resemblance pervading a whole department.[35]

Prichard appropriated his analytical framework for classifying humans from eighteenth-century naturalists. In particular, he referred to the work of Blumenbach, Cuvier, Linnaeus and George-Louis Leclerc Comte de Buffon (1707–88). Each of these figures provided him with varying techniques for ordering his data. Linnaeus's method for arranging specimens according to physical characteristics was fundamental because it afforded observers flexibility when organizing their material. A variety could continually be modified within the Linnaean system depending on how the observer interpreted the evidence collected. If a new specimen possessed a characteristic that was not previously attributed to a variety, it could be added to its group's description. A previously unrecorded feature did not necessarily denote a new distinct species. Prichard stated that 'Of this kind are the genera of the older zoologists, as well as the classes and orders in the several systems of Linnaeus and other botanists, and natural historians. The principle of these arrangements being an arbitrary definition, they may be changed or modified ad infinitum according to the caprice of the constructor'.[36]

Unlike varieties, species remained constant, and Prichard argued that 'it is not the case of species. Here the distinction is formed by nature and the definition must be constant and uniform'.[37] A species could only reproduce with its own kind under this model. Therefore, because varieties were deviations within the same group, they could intermix and propagate to form offspring that were 'prolific', meaning fertile. The ability to produce prolific progeny was a fundamental quality of a species. Prichard wrote,

> if a male and female produce an offspring which is prolific, the tribes to which the parents respectively belong, are hence proved not to be specifically different, and whatever diversities may happen to characterize them, are in this case looked upon as examples of variation. But if the third animal be unprolific, it is to be concluded that the races from which it is descended, are originally separate or of different kinds.[38]

Under this model, if the third generation of progeny were able to produce offspring, they were deemed variations of the same species. Prichard utilized natural history terminology and classification methods in his ethnological paradigm to support his monogenetic theory, and this was an essential component of his observational practice. By using a taxonomy that focused on the physical features of different races, Prichard aimed to demonstrate that variations in human form were examples of varieties, not of species. He claimed that his framework provided observable evidence for the unity of races, and this in turn would further strengthen his suppositions.

## *Lawrence, Natural History Classifications and Monogenism*

Comparative studies were also fundamental in Lawrence's method for studying human variation, and he incorporated natural history observational techniques into his ethnological model. He wrote for instance that practitioners were 'to compare or contrast with each other the various nations or tribes of human beings; to delineate the physical and moral characters of the people inhabiting the different portions of the globe, and to trace their progress from the first rudiments of civil society to the state at which they now arrived'.[39] Lawrence continued by arguing that having an extensive knowledge of several research fields was helpful when tracing the origin of humans because each analytical tradition provided different insights into the past. He wrote,

> The anatomist and physiologist unfold the construction and uses of the corporeal mechanism; the surgeon and physician describe its diseases; while the metaphysician and moralist employ themselves with those functions which constitute the mind, and with the moral sentiments. Man in society, his progress in the various countries and ages of the world, his multiplication and extension, are the province of the historian and political economist.[40]

This passage is significant because Lawrence believed most studies written on human diversity were limited in scope.[41] Accordingly, he contended that natural histories regarding humans were too narrow because they failed to incorporate the methodologies of scholars from other fields of inquiry. Thus, for example, political economists did not consider the observations of anatomists, and vice versa. Responding to this, Lawrence wanted to produce an integrative model for studying races which combined techniques from any research field that could shed light on human variation. He argued, 'To write such a history of our species would demand a familiar acquaintance with nearly the whole circle of human knowledge, and a combination of the most opposite pursuits and talents'.[42]

Lawrence also wanted to demonstrate that all humans were of the same species, and he formulated a taxonomical system for ordering and interpreting humans as varieties. An entire section of his book was devoted 'to the different races of mankind ... [with] discussion of the characters by which they are distinguished, and full consideration of the question, whether they ought to be regarded as originally distinct species, or as varieties of one single species'.[43] He continued by stating that 'we may suppose that one kind of human beings only was formed in the first instance; and account for the diversity, which is now observable, by the agency of the various physical and moral causes to which they have been subsequently exposed; in which case they will only be for different *varieties* of the same species'.[44] Positioning different races as varieties within a single species was centrally important for Lawrence's ethnological model, and he believed that such a method provided observable evidence for monogenism.

His training in medicine was fundamentally important because he viewed a combination of natural history and anatomical and physiological comparative studies as the cornerstone of ethnological research. He stated that

> It will be necessary to ascertain carefully all the differences that actually exist between the various races of men; to compare these with the diversities observed ... to apply to them all lights, which human and comparative physiology can supply; and to draw our inferences concerning their nature and causes, from all the direct information, and all the analogies, which these considerations may unfold.[45]

However, Lawrence did not want to form a taxonomical framework that just focused on the differences exhibited between human races; he also wanted to demonstrate their common links and position human varieties as gradations of the same species. Accordingly he argued, 'It is necessary to examine, not only the more marked differences, but also the numerous gradations by which opposite extremes are in all cases connected and gradually brought together: it is also necessary ... to compare with man the various living beings which more nearly resemble him. The whole proceeding must be governed by the principles of general physiology'.[46]

A clearly defined explanation for the terms 'species' and 'varieties' was important because if practitioners learning to study humans could not identify the difference between the two categories, Lawrence's taxonomical framework would be inadequate. He wrote, 'Before entering on the immediate object of this section, it is necessary to consider what is the precise acceptation of the terms *species* and *variety* ... [and] what constitutes a species and how varieties arise out of it'.[47] As with Prichard's *Researches*, a primary characteristic of a species was that they could produce offspring that were prolific.[48] Lawrence agreed with this point, and in addition he argued that two specimens from separate species would never procreate because of a natural deep-seated sexual aversion towards other species. He wrote,

> Animals are characterized by fixed and definite external forms, which are transmitted and perpetuated by generation. The offspring of sexual unions is marked with all the bodily characters of the parents. However strong the impulse may be, which leads to the continuation of the species, there seems to be an equally powerful aversion to intercourse with those of other species.[49]

Lawrence argued that logically, when two beings of differing physical appearance propagated to produce offspring, they were not separate species but rather different varieties. The uniformity of characteristics in species was maintained because of a natural inherent aversion to other kinds.[50] Accordingly, for Lawrence, other factors produced varieties in a species. For example, he wrote, 'Variation in the quantity and quality of food may cause some slight differences'.[51]

Lawrence continued by linking humans to the rest of the animal kingdom and argued that the same classificatory rules for different species of animals should be applied to humanity. He stated, 'This method is the only satisfactory one of investigating the varieties of the human species. The diversities of physical and moral endowment which characterize the various races of man, must be analogous in their nature, causes, and origin, to those which are observed in the rest of the animal creation, and must therefore be explained on the same principles'.[52] Lawrence argued that researchers had to consider 'the circumstances in which the several races of men differ from each other, [and] to compare them to the corresponding differences of animals'. In doing so, he concluded, 'the particular and general results of these inquiries lead us plainly to the conclusion, that the various races of human beings are only to be regarded as varieties of a single species'.[53]

For both Prichard and Lawrence, natural history taxonomical systems were centrally important for their ethnological observational practices. Natural history provided them with an analytical framework that explained to their readers how classification systems worked and their utility for studying human diversity. We have seen how Prichard and Lawrence organized races according to their physical features and separated them into varieties within a single species. In doing so, both figures believed they were providing their readers with empirical evidence for monogenism. However, the argument did not stop there, and once Prichard and Lawrence provided a general theoretical and practical methodology for classifying humans, the next step was to further refine their observational mode by incorporating their anatomical – and to a lesser degree their physiological – training into their ethnological paradigms.

## Ethnology and Medicine

> It is only of late years, and principally through the labours, the lectures, and the excellent writings of BLUMENBACH, that the natural history of man has begun to receive its due share of attention ...[54]
>
> William Lawrence, 1819

### *Prichard's Physical Descriptions of Races*

The ability to differentiate both pronounced and subtle variations in the physical structure of races was an important skill for early ethnologists. Prichard's training in comparative anatomy – and to some extent physiology – influenced his taxonomical model, and one of his primary aims was to develop and refine a method for analysing the skin, eye, hair and skull types of different racial groups. Building upon Blumenbach's classificatory framework in *De Generis Humani Varietate Nativa* (1775), Prichard provided his readers with explanatory definitions for each of his seven racial varieties. The purpose of these descriptions was to outline the main distinguishing characteristics to look for when ordering humans into taxonomical groupings. Prichard also provided geographical regions for each type. His first group were albinos, and he stated:

> The most prominent peculiarities of this class of men are the following: The iris of the eye is of a bright red hue, and the organ of sight remarkably light. The complexion is either uncommonly fair, and resembling that of the most exquisite examples of the sanguineous temperament, or it is of a dull whiteness of disagreeable aspect, and giving the appearance of disease. The hair is extremely soft in its texture, and in general is perfectly white, but in some instances of a very light flaxen colour, and when this variety springs up among Negroes, the woolly excrescence which cover the heads of that race is white.[55]

Prichard explicated for his readers what he considered to be the essential physical characteristics of an albino, including their skin, eye and hair colour. Bringing together each of these aspects, the aim was to teach his readers how to observe an individual from this group in a specialized way. Following his description, a novice observer learned how all the components of an albino's structure came together to create a conception of this variety as a whole. Interestingly, Prichard noted that albinos were located in multiple regions of the world.

Prichard constructed each of his other racial varieties in a similar fashion. For example, Prichard's second and third groups represented the two main physical appearances of Europeans. First there was the 'yellow haired variety', and he wrote that '[this] ... variety of human complexion is marked by hair of a reddish, yellowish, or flaxen colour, and a skin very fair, though not so white as that of the last mentioned description of men [the albinos]'.[56] He continued by stating that 'The iris of the eye is always of a light hue, generally blue or grey, the shade of colour bearing a relation to that of the hair and skin'.[57] Prichard associated this likeness of humans to central Europe, where the Germanic tribes originated. He did so because many of the inhabitants of that region of Europe fit his description.[58]

Prichard's third group represented Europeans of a darker complexion, and he associated this likeness with the Mediterranean region of Europe. He stated that this variety of humans was 'distinguished by dark or black hair, with the iris of a corresponding hue, while the complexion is white, though without that delicate tint which characterizes the sanguineous constitution. The skin soon becomes brown by exposure to the sun, but in persons who are constantly protected from the influence of the weather, it is frequently almost of the whiteness of marble'.[59] Prichard argued that the colour of a person's hair and iris had a correlation to the darkness of their skin colour. For instance, a person with light-coloured hair usually had blue or grey pigmentation in their iris and fair-coloured skin.

If a person had a darker skin complexion, however, they usually had brown or black eyes and hair. According to Prichard, this relationship between eye, hair and skin colour was most visible in his four other racial varieties. Prichard's fourth group inhabited eastern Asia, in countries such as China and Singapore. He wrote that this variety had 'A complexion of a yellowish tint passing into an olive, and stiff long black hair'.[60] His fifth group consisted of inhabitants from North and South America, and he described these people as 'a class, which is characterized by a complexion darker than the preceding [fourth variety]'. He added that their skin pigmentation varied from 'a copper colour to a more dusky hue' and that they had black hair and brown eyes, a consequence of their darker complexion.[61]

Prichard's sixth variety consisted of inhabitants from Africa. However, unlike his other descriptions of racial varieties, Prichard discussed a peculiar characteristic that occurred with members of this group. He wrote, 'The children of Negro parents are sometimes variegated, having their skin diversified with black and white spots, and part of their woolly hair white. They are commonly called pie-bald Negroes. This variety is not very rare in the West Indies ... The white spots have the same hue as the skin of a very fair European'.[62] The purpose of describing this anomaly was to demonstrate an exceptional instance where the shade of a subject's hair did not entirely correlate with the colour of their skin. Prichard's final variety consisted of people from central Asia and parts of the South Pacific. He noted that a 'Black or dark tawny colour forms the complexion of several races of men'.[63]

Prichard combined his knowledge of natural history classification systems and his basic training in anatomy in order to further refine his taxonomical system of humans. He argued that the differences in physical appearance between his seven racial varieties were for the most part superficial, and they were comparable to the differences that existed between varieties within any animal species. Therefore, humans could be classed under one group. He wrote,

> Such are the varieties of colour observed in the human kind. They are clearly shown by the foregoing comparison to be phænomena analogous to the deviations which continually occur in the inferior species of animals. We are therefore compelled, according to the received laws of reasoning on physical questions, to refer the former to the same class of natural appearances with the latter. It may be concluded, that in the various colours of men there is certainly no specific difference.[64]

Prichard proceeded by shifting his focus to a discussion on how physical characteristics were passed down to offspring, and he described this process as the 'hereditary transmission of the varieties'. Essentially he argued that children acquired the physical features of their parents. Thus if both parents belong to Prichard's third variety, their child would most likely have a light complexion with dark hair and eyes.[65] However, in many cases where two people from different varieties procreated and had children, Prichard stated that the appearance of their progeny would resemble only one of the parents' physical likenesses. He wrote, 'The offspring of parents of different varieties sometimes partakes of the characters of both, and forms an intermediate class; in other instances, the mixed progeny resembles either the father or mother distinctively'.[66]

Prichard also discussed the issue of diversity in nature in order to strengthen his monogenetic theory. Because he argued that all humans came from a single origin, he had to account for why the various races looked different. He did so by arguing that the concept of variation in itself was a universal feature, and he stated that 'we perceive no common quality so universally characterising the works of Nature, as an infinite and inexhaustible [as] variety'.[67] For Prichard variation in race was no different to variation among siblings. He argued that 'deviation exists even among the individuals of the same family'.[68] Thus Prichard believed that if there was a divergence in the physical appearance of members of the same family and they were from the same stock, racial groups could appear physically different and still come from a single origin.[69]

Building further on his basic taxonomical system of human varieties, Prichard shifted his attention to a more specialized examination of the physiognomy of human skulls. He described three main structural types and wrote, 'The most striking, and important instance of diversity in the human form, is in the configuration of the skull. Physiologists have directed much of their attention to this variety, and have invented several methods of classifying the peculiar appearances, and reducing them to general principles'.[70] Accuracy of observations was essential, and Prichard utilized techniques from three continental anatomists: Blumenbach, Cuvier and the Dutch physician Pieter Camper (1722–89). Prichard would have learned all of these modi operandi during his days studying anatomy at Edinburgh. Nevertheless, he preferred Blumenbach's method for studying skulls outlined in *De Generis Humani Varietate Nativa*, and he wrote, 'Blumenbach's mode of examination affords a much more ample and accurate view of the diversities of the cranium. It regards chiefly the form of the frontal and superior maxilliary bones, as giving the most important characters, and those on which the general description of the head principally depends'.[71]

When Prichard published the first edition of his *Researches*, he was a fairly unknown figure within the scientific and medical community. His use of Blumenbach's observational technique for describing skulls added some authority and credibility to his work; after all, Blumenbach was one of the most eminent anatomists in Europe. His classificatory system had five skull types (see Figure 1.1), but Prichard modified the system slightly by changing it to three main skull configurations and two additional subtypes.

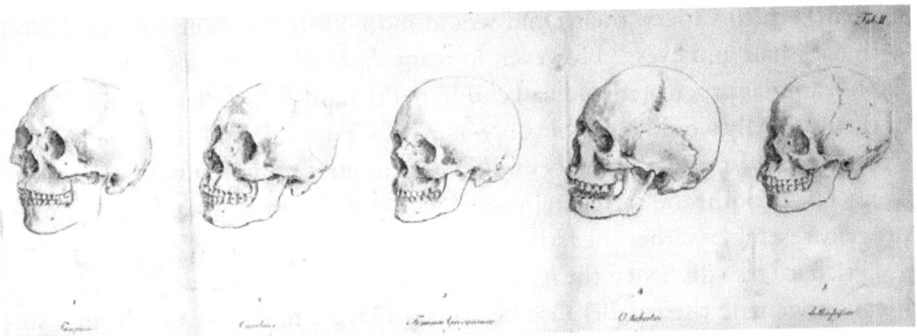

**Figure 1.1: Examples of Blumenbach's five skull conformations. From J. F. Blumenbach, *De Generis Humani Varietate Nativa*, 3rd edn (Göttingen: University of Göttingen, 1795), Tab. II. Reproduced by permission of the Wellcome Library.**

The system gave specific instructions on how to set up the skulls for observation, and this was done to promote a standardized technique, so that others could reproduce Blumenbach's original results from the 1770s.[72] Prichard wanted his readers to learn the method, so he translated Blumenbach's instructions and descriptions into English for his readers. He wrote,

> [Blumenbach] placed the skulls he wished to compare in a row on a table, together with the lower jaws, in such a manner, that the cheek bones or zygomatic processes should all touch the same horizontal line. Then directing his view to the vertex of each, he described all the peculiarities apparent on the front which it thus presented, and this method he called his vertical rule.[73]

The first skull type within Blumenbach's taxonomy was that of the 'European', and he described it as follows: 'The head is of the most symmetrical form, almost round; the forehead of moderate extent; the cheek bones rather narrow, without any projection, but having directions downwards from the malar process of the frontal bone; the alveolar edge round; the front teeth of either jaw placed perpendicularly'.[74] Blumenbach carefully depicted the main physical features to look for when examining, recording and classifying a skull. He noted both its general physical structure, such as its 'symmetrical form', and its specific characteristics, such as the 'narrowness of its cheek bones'.

Blumenbach's second skull type – which he called the 'Mongole' – was similarly described, and he stated that 'The head is almost square, the cheek bones projecting *outwards*; the nose flat; the space between the eye-brows, and nasal bones nearly in the same horizontal plane with cheek bones; the supercilliary arches scarcely to be perceived; the nostrils narrow; the fossa maxillaris slightly marked; the alveolar edge in some degree rounded forwards; the chin slightly prominent'.[75] Having clearly defined descriptions of different skull types imparted the observer with the ability to conduct comparative anatomical analyses. For instance, in the case of the Mongole skull, Blumenbach depicted its general form as being 'almost square', and this contrasted significantly with his description of the European's 'symmetrical' shape.

The final skull type in Prichard's modified version of Blumenbach's taxonomy was the 'Negro' configuration, which he described as 'narrow, compressed at the sides; the forehead very convex, vaulted; the cheek-bones projecting *forwards*; the nostrils wide; the fossæ maxillares deeply marked behind the infra-orbital foramen; the jaws lengthened; the alveolar edge narrow, long, and elliptical; the front teeth of the upper jaw turned obliquely forwards; the lower jaw strong and large; the skull in general thick and heavy'.[76] As was the case of the first two skull types, Blumenbach's description identified both the general shape of the skull and its specific features. Prichard stated that 'Such is the description of the heads of the three races of mankind, which differ most widely from each other'.[77] There were two additional skull configurations in Blumenbach's classificatory system: the 'American' and the 'Malay'.[78] However, Prichard believed that these skull configurations were subtypes because they possessed significant overlapping characteristics with the other conformations. Thus, using his anatomical training, Prichard discussed both basic physical features in different races, such as their skin, eye and hair colour, and more specialized characteristics, such as the configurations of skulls. In doing so, he aimed to further support his monogenetic theory with observable evidence.

## *Lawrence's Physical Descriptions of Races*

Lawrence's ethnological framework also looked at different races and divided humans into groupings based on skin, eye and hair colour. He wrote that 'the different tints and shades of the skin, offering themselves so immediately to observation, and forcing themselves, in a manner, on the attention of the most incurious have always been regarded by the generality of mankind as the most characteristic attribute of the various races'.[79] However, according to Lawrence, many of the racial descriptions recorded by observers interested in human diversity often misled the reader because of the informative language used. He stated,

> The very nature of language, the want of adequate expressions to denote the endless shades of colour, and the indeterminateness of those which are applied to various tints, create some difficulties in this part of the subject, by producing considerable discrepancies in the reports of travellers, which again are of course increased in many cases by haste and carelessness; by superficial examination, and loose choice of expressions.[80]

Lawrence wanted to create a taxonomical system that divided humans into six different varieties, which were clearly defined by their chief anatomical characteristics and their geographic locality. Observers would thus have a standardized lexicon to utilize when ordering humans into different groupings. Lawrence identified his first racial variety as 'white', and he noted that 'redness of cheeks is almost wholly confined [to this group], being observed, at all events, very rarely in the other varieties'. Accordingly, the ability to blush was a demarcating characteristic for members of the 'white' variety. However, blushing was not an exclusively 'white' feature, and Lawrence did provide examples of this phenomena among people of fairer complexions from other racial groupings. He stated for example that 'in some very light examples of the brown and yellow varieties, blushing has been noticed'.[81]

Lawrence's second racial variety were the 'albinos', and he described this group as having 'a skin of a peculiar reddish, or an unnatural white tint, with corresponding yellowish white or milk white hair, and red or at least very light blue or grey eyes ... The hair of all parts of the body is unnaturally white and soft; it has not the snowy whiteness of old age, nor the elegant light yellow or flaxen appearance of the fair-haired'.[82] Interestingly, Lawrence's didactic explanation of albinos' chief characteristics was similar to Prichard's account from the *Researches* – both Prichard and Lawrence described albinos as having very light skin, eye and hair colouring.[83] Lawrence identified his third racial group as 'yellow or olive', and he wrote that members from this variety had 'a middle tint, between that of a ripe wheat and boiled quince or dried lemon-peel'.[84] This depiction characterized the 'Mongolian tribes, usually called, together with the inhabitants of great part of Asia, [and the] Tartars'.[85] Situating the geographical region for each racial type was significant because if an observer was struggling to differentiate and categorize a subject based on their physical form, they could still class them into a variety based on their locality.

This was particularly problematic with Lawrence's fourth and fifth racial groupings because there were several overlapping features between the two types. A novice observer could easily struggle to identify the differences between them. For instance, Lawrence's fourth designation was described as 'Red or copper colour', and he stated that members from this group possessed 'an obscure orange or rusty iron colour, not unlike the bark of the cinnamon-tree'.[86] He continued by noting that members from this taxonomic group inhabited 'nearly the whole continent of America, and is almost confined to that division of the globe'.[87] Conversely, Lawrence described his fifth variety as 'Brown or tawny', and their skin pigmentation as 'of a middle tint between the colour of fresh mahogany and of cloves or chestnut'. Geographically, however, he placed his fifth racial group in 'the Malays, and most of the inhabitants of the numerous islands scattered through the Pacific Ocean'.[88] By comparing these two descriptions, we can clearly see similarities within both accounts, and travellers using Lawrence's text could easily confuse a 'mahogany'-coloured skin tone with that of the 'bark of a cinnamon tree'. Thus providing readers with geographic locations for each variety helped to minimize misidentifications.

Lawrence identified his final group as 'Black', and he noted that members from this variety of humans were of 'various shades, from the sooty colour or tawny black, to that of pitch or ebony, or jet black'. Moreover, Lawrence argued that people from this racial grouping inhabited

> the continent of Africa, characterizing all the Negro tribes. It is found also in the Negro-like natives of New Holland, Van Diemen's Land, Papua or New Guinea, the New Hebrides, and other islands of the South Sea; and is seen, mingled with the national colour, in Brazil, California, and India. The New Caledonians constitute an insensible transition, with the chestnut-coloured islanders of Tongataboo, and the dark New Hollanders, from the tawny or brown Otaheiteans to the Papuas or Negroes of New Guinea.[89]

For Lawrence, it was important to sketch out his six main varieties of colour while at the same time ensuring that they were not overly concrete. As we saw with Linnaeus's classifications, descriptions of varieties were flexible and allowed the researcher to continually modify and add features depending on what was observed in nature. By showing the fluidity of boundaries between the different varieties, a naturalist could demonstrate overlapping characteristics as evidence for a single species.

Lawrence argued that there were 'intermediate shades' that existed between different varieties, and he stated that 'In describing these varieties, it is necessary to fix on the most strongly-marked tints, between which there is very conceivable intermediate shade of colour. The opposite extremes run into each other by the nicest and most delicate gradations; and it is the same in every other particular, in which the various tribes of human species differ'.[90] Lawrence wanted to strengthen his case for monogenism by creating a taxonomy based on physical features, which organized humans into adjustable groupings. He stated that such a method formed 'no slight objection to the hypothesis of distinct species: for, on that supposition, we cannot define their number, nor draw out the boundaries that divide them'.[91]

As we have seen, Prichard argued that physical features of parents were 'transmitted' to their progeny.[92] However, if the parents of a child were from different varieties, the appearance of the offspring would either comprise a mixture of characteristics from both parents, or only the physical properties of a single progenitor. Lawrence had a similar theory, and he wrote that 'When two varieties copulate together, the offspring resembles neither parent wholly, but partakes of the form and other properties of both. This cannot with propriety be termed *hybrid* generation'.[93] To assist his readers with accurately identifying the progeny of mixed-racial parentages, Lawrence included a graph (see Figure 1.2) in his book that listed the 'appropriate' designations to use when describing a person with parents from different racial varieties.

* The proportions are represented as follow, according to the principles sanctioned by usage:—

| *Parents.* | *Offspring.* | *Degree of Mixture.* | |
|---|---|---|---|
| Negro and European | Mulatto | ½ white | ½ black. |
| European and Mulatto | Terceron | ¾ —— | ¼ —— |
| Negro and Mulatto | Griffo, or Zambo | ¾ black | ¼ white. |
| European and Terceron | Quarteron | ⅞ white | ⅛ black. |
| Negro and Terceron | | ⅞ black | ⅛ white. |
| European and Quarteron | Quinterin | $\frac{15}{16}$ white | $\frac{1}{16}$ black. |
| Negro and Quarteron | | $\frac{15}{16}$ black | $\frac{1}{16}$ white. |

The two latter are respectively white and black; and of these, the first are white by law, and consequently free, in our West-India Islands. All remains of color are so completely banished, that they are not distinguishable from whites in any respect.

Figure 1.2: Lawrence's graph for 'accurately' identifying people of mixed racial descent. From W. Lawrence, *Lectures on Physiology, Zoology, and the Natural History of Man, Delivered at the Royal College of Surgeons* (London: Foote and Brown, 1828), p. 259. Reproduced from the author's private collection.

Lawrence sought to further refine the quality of observations written by researchers in Britain and by informants travelling abroad. Illustrative graphs such as the example in Figure 1.2 demonstrate that he wanted to decrease the amount of discrepancies recorded by observers of human diversity, by formulating a clear method for identifying and ordering racial varieties and bi-racial varieties. He based his methodological model on his training in anatomy and natural history.

Prichard argued that people with darker skin complexions had dark eyes and hair colour. Lawrence also observed this characteristic within his taxonomical framework, and he further noted that 'The four coloured varieties of men have black hair, which is always stronger and coarse in texture than in the whites. This difference is particularly noticed by the Chinese, who contemptuously compare the hair of Europeans to the soft fur of the smaller animals. In Negroes, native Americans, and New-Zealanders, I have found the texture much stronger than in the darkest Europeans'.[94] Lawrence's descriptions of the four primary hair textures further refined Prichard's classificatory paradigm in the *Researches*, and in doing so, it aimed to help readers identify and organize people into different varieties.[95] He also provided descriptions of each type for his readers to utilize in their studies.

Lawrence described his first hair grouping as 'Brownish, deviating into yellow (flaxen) or red on one side, and black on the other; copious, soft, long, and forming more or less distinct ringlets or undulations. It is seen in the temperate climates of Europe'.[96] This group would include people from the Netherlands, Scandinavia and Germany. Lawrence hoped that by recognizing which regions particular hair types were commonly observed in, his readers would more accurately classify populations. Lawrence stated that his second hair type – which he described as 'Black, strong, and thin' – was commonly seen in 'the Mongolian and American varieties'.[97] This was in contrast to the African variety, which Lawrence described as 'Black and crisp, so as generally to be called woolly; [and] common to all the Negro tribes'.[98] Finally, Lawrence concluded his discussion on hair colouration by stating that 'a similar connexion in point of colour to that which I have just explained between the skin and the hair exists also between the former organ and the eyes'.[99]

Lawrence also discussed different facial structures in his lectures on the *Natural History of Man*, and he built his examination on Blumenbach's taxonomy from the 1770s. Lawrence stated that 'The five following varieties are established by Blumenbach after a careful comparison of numerous drawings, and of the various races themselves, in situations where commerce attracts them from all parts of the globe'.[100] Interestingly, Lawrence focused on the facial conformations of different varieties, and this examination varied in some respects from Prichard's analysis on skull types. However, because both scholars used Blumenbach's work as the foundation of their respective taxonomies, there was some significant overlap. The first group that Lawrence considered was Blumenbach's 'European' facial type. He wrote that the European possessed

> An oval and straight face, with the different parts moderately distinct from each other; high and expanded; nose narrow, and slightly aquiline, or at least with the bridge somewhat convex; no prominence of the cheek-bones; small mouth, with lips slightly tuned out particularly the lower one; a full and rounded chin.[101]

Lawrence continued by remarking that such an appearance was in accordance with general conceptions of beauty.[102] Blumenbach's second facial type was the 'Mongole', and Lawrence described this group as

> Broad and flattened face, with the parts slightly distinguished, and as if it were running: the space between the eyes flat and very broad, flat nose, rounded projecting cheeks: narrow and linear aperture of the eyelids extending towards the temples ... the internal angle of the eye depressed towards the nose, and the superior eyelid continued at that part into the inferior by a rounded sweep; chin slightly prominent.[103]

Using descriptions such as Blumenbach's depiction of 'Mongole' faces were important because they provided readers with concise representations of the primary racial groups inhabiting the world. He hoped that these descriptions would encourage the standardization of data on human varieties because British researchers analysing travel reports, and informants overseas, would know exactly what sorts of characteristics to record when observing different regional populations.

The third facial variety was commonly observed among the inhabitants of the Americas, and Blumenbach identified these populations as having a 'Face broad but not flat and depressed, with prominent cheek-bones; and the parts, when viewed in profile, as it were more deeply and distinctly carved out. Short forehead, eyes deeply seated, nose flattish, but prominent'.[104] Lawrence also included Blumenbach's other two facial configurations, the 'African' and the 'Malay'. Overall, each description provided readers with both the general features of the face, such as the 'broad but not flat' conformation of the 'Mongole', and specific qualities, such as the 'very thick upper lips' of the 'African'.[105]

For both Prichard and Lawrence, a combination of natural history and anatomy was critically important for their ethnological observational practices. With the examples provided, we can see that both scholars utilized their training in medicine and natural history to organize humans into types based on their eye, hair and skin colour. Moreover, they based their models on Blumenbach's taxonomy. In doing so, Prichard and Lawrence believed they were providing further evidence for monogenism. However, in order for these figures to analyse the anatomy and physiology of races and organize them into taxonomic groupings, it was essential to collect data. For this, Prichard and Lawrence used the accounts of voyagers.

## Ethnology and Voyages of Exploration

> The Men are here more numerous than at *Dusky Bay* ... They are all a stout, well limbed race of people in the upper part. The knees large, the legs small & bandy, caused by their way of sitting. They are of a swarthy complexion, have all black, somewhat curled hair, which they, when dressed tie up on the crown of the head ...[106]
>
> Johann Reinhold Forster, 1778

### *Prichard and Travelogues*

Prichard was tied to England because of his professional commitments in Bristol and London. Thus he relied on the reports of scholars from various fields of inquiry, as well as travellers, for his data. However, he was not simply anthologizing the material that he collected. Rather, he organized the evidence into smaller and more manageable groupings so that he could analyse and interpret its meaning. He was highly attuned to problems associated with using second-hand sources, and he aimed to minimize the amount of discrepancies they produced through extensive cross-comparative studies. The meticulous nature of his work was significance because Prichard was attempting to establish himself as an authority within the emerging research field.

Looking at the footnotes from the first edition of his *Researches*, we can see that Prichard's evidence came from a broad range of different sources.[107] His fluency in various languages made it possible to incorporate data from ancient Greek and Latin texts as well as more recent works from continental writers of the late eighteenth and early nineteenth century. Thus Prichard drew upon classic Greek philosophers such as Aristotle (384–322 BCE), Heraclitus of Ephesus (*c.* 540–475 BCE), Plato (429–347 BCE) and the dramatist Sophocles (495–*c.* 406 BCE) for information on the early existence of humans. Moreover, he included travel descriptions from explorers such as James Cook (1728–79), Scottish voyager James Bruce (1730–94) and French circumnavigator Louis Antoine Bougainville (1729–1811). Prichard also cited the works of the French orientalist Joseph de Guigne (1721–1800) and the Russian naturalist and physician Johann Güldenstadt (1745–81).[108]

Once Prichard read the writings of various scholars and travellers, his next task was to bring together into a narrative a comprehensible description of both the physical and cultural characteristics of different races from around the world. He did so by first outlining a methodology for analysing secondary sources. He wrote,

> Whenever the testimony of historians affords any information, we shall avail ourselves of this resource. But the direct authority of history furnishes but a very imperfect insight into the origins of nations. We must therefore often depend on the reflected light which is obtained by the comparison of languages, by the analysis of civil and religious institutions and mythological fables, or by tracing clearly marked affinities in the manners and customs of different tribes.[109]

Prichard explained his comparative method of analysis and demonstrated for his readers what sorts of evidence to consider when examining different human varieties. In addition, he linked this technique to his belief in monogenism. For Prichard, if there were similar languages or cultural practices occurring within two distinct indigenous communities, this commonality potentially meant that the two groups shared a common origin. This was further strengthened if the two populations were geographically distant from each other, with no evidence to suggest that there was any interaction between them. He argued,

> The most important of these aids is the comparison of languages. When two nations ever so distantly separated, or however widely distinguished in all other points, are found to speak the same language, or to use dialects which though differing in pronunciation and otherwise variously modified, can yet be traced to the same radicals or elements, so as to prove that an essential affinity existed in their primitive structure, it is certain that such nations have descended in great part from the same stock.[110]

Prichard recognized that such a method was fallible, and he stated that 'This argument is certainly liable to be abused. Many words may be introduced in the speech of any tribe by their immediate neighbours, if frequent communication subsist between them'.[111] However, with careful analysis, he believed such a theoretical framework could produce illuminating results. Prichard wrote that 'it is possible to discover, by proper discrimination, what parts of the vocabulary are thus adventitious, and what are radical, elementary, or original'.[112] Under this model, a skilled researcher with a background in philology and a proficient knowledge of languages would be able to discern – with a trained eye – the important linguistic characteristics to look for when seeking to connect races through their lexicons. A researcher could also apply this discriminating technique to other sorts of cultural phenomena such as religious practices or mythology.

Michael Bravo argued that 'For the ethnologicals ... language held the key to unravelling the story of humanity. Although comparative studies of evidence such as skulls and other remains could illuminate the effects of climate on the physical history of the races, these could only serve as a subsidiary criterion to that of language'.[113] By reading the descriptions of travellers such as Cook and Bougainville, Prichard looked for shared linguistic and physical characteristics in an attempt to bring together different ethnic groups and show their common origin. When discussing another South Pacific island of New Caledonia, Prichard wrote that 'The people of New Caledonia have some customs similar to those of the Tannese and Mallicollese ... Their language resembles that of Tanna of New Zealand, and the Friendly Isles'.[114]

It was essential for researchers writing about human diversity to be scrupulous with their observations, and accuracy in the descriptions of races was of the utmost importance. Prichard wrote that 'in this inquiry caution and accuracy are requisite'.[115] One of the chief problems, according to Prichard, was that it was difficult to be exacting when observing extra-European cultural habits and customs. This was because it was easy for the traveller to misjudge the significance of cultural practices, which differed significantly from their own Western explanatory models of the world. Moreover, if a traveller examining indigenous peoples *in situ* observed similarities, this was again to be interpreted as evidence in favour of common origin. Prichard stated,

> Those shades of character and manners which have their origin in the general principles of human nature, or arise from circumstances and situations likely to occur to all men, may be found to prevail more or less among tribes which have had no intercourse. But if we find clear coincidences in such peculiar habits and customs as are purely arbitrary and casual, we cannot suppose these instances to have been of separate production, but are compelled to acknowledge that they evince a common origin or a connexion at some former period between the nations who continue to be marked by such traits.[116]

Once Prichard sketched out his interpretive framework for connecting and analysing the observations of travellers, he then shifted his focus to the different regions of the world and the races that inhabited them.

It is beyond the scope of this study to examine each of the geographic locations covered in Prichard's *Researches*. It is sufficient, for our purposes, to focus specifically on his exemplary examination of the South Pacific and its inhabitants. Under the heading of a 'General view of the nations inhabiting the South Sea Islands, and the Austral Countries', Prichard discussed in general terms the climate, flora, fauna and peoples of the region. Because the South Pacific had an incredibly diverse environment, Prichard believed that it afforded researchers with an opportunity to observe the effect various external influences had on humans. He wrote, 'These insular countries are distributed through almost every variety of climate, and contain abundant diversity of local situations; therefore they afford us an opportunity of observing whatever influence physical causes may be supposed to exert over our species'.[117]

The remoteness of the South Pacific was another important issue for Prichard. Because many Pacific islanders remained fairly untouched by European contact and settlement, he believed that it provided researchers with a glimpse into the early history of humanity. He continued by stating that

> we also derive advantage from the remote distances which separate the islands, and from the imperfect knowledge of navigation which the natives possess: for these circumstances prevent intercourse among the different tribes, and preclude those frequent changes or intermixtures of population, which perplex our inquiries into the history of continental nations.[118]

After outlining the general features of the South Pacific, Prichard's next step was to transform travelogue accounts into ethnographic observations. Prichard was providing his readership with detailed information on extra-European peoples, which for many of his readers was their first 'encounter' with indigenous communities.

When describing a native population, Prichard pieced together the reports of several different explorers with the use of both quotations and general reference to their works. This was significant because it was an attempt to authenticate his racial depictions. The only way Prichard could acquire direct experience observing extra-Europeans *in situ* was by visiting foreign people shows in London. However, there is little evidence to suggest that he attended any of these exhibits. For instance, in the 1830s Prichard was unable to accept a rare opportunity to meet a Mandingo performer in the British capital.[119] Thus referring to the great travelogues of explorers such as Cook or Bruce legitimized his analysis as a verisimilar representation based on eyewitness accounts. Accordingly, Prichard wrote, 'we must follow in detail the observations of voyagers'.[120]

Prichard's use of travel descriptions was part of a four-step process. Within this framework, he saw travellers as data collectors and their observations as types of natural history specimens. He attempted to establish the credibility of these reports by emphasizing the accuracy of their observations. Once this was done, he would classify this evidence into groupings according to the similarities and differences described in the accounts. The final step within the process was to provide an analysis of the material and identify the significance of certain key passages from the travel narratives. Prichard recognized certain explorers as having particularly significant reports. For instance, he wrote, 'The different voyagers who preceded Captain Cook in exploring the Pacific Ocean had given us many curious notices concerning the natives of the islands, but we have derived more extensive information from the remarks of that celebrated navigator [Cook] and the naturalists who accompanied him, among whom Dr. Forster holds a distinguished place'.[121] Prichard gave particular weight to the observations of Johann Reinhold Forster (1729–98) – the naturalist from Cook's second and third voyages – because of his background in both natural history and medicine. For Prichard, Forster's travelogue was exceptional because he was trained in two of the traditions fundamentally important for ethnological research.

Prichard also identified the observations of the physician and naturalist William Anderson from Cook's second and third voyages as reliable because of the detail and accuracy of his reporting. Prichard wrote, 'For the most judicious and accurate accounts we possess, we are indebted to Mr. Anderson, who seems to have combined more of the qualities of a philosophical observer than almost any other individual who has traversed the regions in question'.[122] He continued by stating that 'As far as his observations extend, we shall form our opinions on his testimony, and whenever his statements differ from the more careless or superficial remarks of others, we shall rely on the former with a well authorized confidence'.[123] Prichard was further impressed that Anderson paid special attention to language in his descriptions, and he noted that

> Mr. Anderson was well aware of the importance of a comparison of languages in tracing the history and connexion of different tribes. He never omitted any opportunities which occurred of collecting vocabularies, in order to institute an investigation of this kind. The general opinion which he was led to form in these inquiries, and which he was assisted by the renour [*sic*] of various observations, he has given us in these words.[124]

Prichard was almost certainly influenced by Anderson's travel writings because the explorer recognized the significance of language for studies on human diversity. Anderson also argued that philological similarities between the lexicons of different indigenous communities denoted the possibility of a common origin. Quoting Anderson, Prichard wrote, '"If we may depend upon the affinity of languages, as a clue to guide us in the origin of nations, I have no doubt, but shall find on a diligent inquiry, and when opportunities offer to collect accurately a sufficient number of these words, and to compare them, that *all the people from New Holland eastward to Easter island have been derived from the same common root*"'.[125]

Once Prichard established the accuracy and authenticity of his travellers' observations, he then correlated their accounts and constructed representations of their physical conformations. In the case of South Pacific peoples, he began by ordering the populations into taxonomical groupings and wrote that 'The primary characters, as nearly as we can discern them, consist in an approximation more or less exact to those of the Africans'.[126] He continued by describing generally their facial structure:

> The facial angle is small, and the forehead low; the head narrow and resembling the Negro in its general conformation; the nose somewhat depressed, but generally full and fleshy towards the point. The structure of the limbs has the same analogy. The legs are long, and the gastrocnemii muscles deficient. This description more particularly applies to the savage class, but in some degree also to the more civilized, and especially to the inferior orders of them.[127]

Prichard attempted to establish the credibility of his analysis by comparing different travelogues to see if their descriptions matched. In other words, he considered a depiction reliable only if the characteristics observed in one travelogue were regularly recorded in other accounts. He argued,

> It will appear more fully, whether with our present knowledge of these tribes their affinity is so decided as to authorize a general description, when we shall have considered separately and compared the most important notices we have concerning the several divisions. This we propose to undertake in detail, and whether the most general conclusion of the authors above quoted shall be established with sufficient evidence or not, we have no doubt of making some interesting deductions from more partial views, which will have all the certainty desirable.[128]

For example, Prichard demonstrated his method by comparing the observations of William Dampier (1651–1715) and Bougainville, who both discussed the inhabitants of New Guinea in their travelogues. Quoting Dampier, Prichard wrote that the people of New Guinea '"were very black, tall and stout people, with great round heads, broad round faces, large bottle noses, and short frizzled hair"'.[129] Prichard continued by stating that 'M. de Bougainville visited the eastern side of New Guinea, and the islands in the vicinity, and to the south east. His description coincides with that of Dampier. He says "the men of New Guinea are black, with frizzled, woolly hair"'.[130] Because both explorers observed 'frizzled woolly hair' in their descriptions, Prichard deemed these credible sources.

Prichard also included large quotations from Dampier in his footnotes to provide his readers with further ethnographic information for their own analyses. The following description of the inhabitants from New Holland was from Prichard's footnotes, and they are identical to the original version in Dampier's travelogue.

> They have great bottle noses, pretty full lips, and wide mouths. The two fore teeth of their upper jaw are wanting in all of them, men and women, old and young: whether they draw them out I know not: neither have they any beards. They are long visaged [*sic*] and of a very unpleasing aspect; having no one great feature in their faces. Their hair is black, short and curled, like that of the negroes: and not long and lank, like that of the common Indians. The colour of their skins both of their faces and the rest of their body is coal black, like that of the Negroes of New Guinea.[131]

Prichard repeatedly emphasized that texts such as Dampier's *Voyages around the World* were reliable because they were meticulously detailed and contained overlapping descriptions with other sources. These books were to be ranked among the best accounts on which to base research. He wrote, 'we have every reason to rely on the veracity and attentive observations of these authors'.[132] This was further strengthened by the fact that figures such as Dampier, Bougainville and Cook were some of the few Europeans to have seen first-hand the South Pacific islands and their inhabitants.

## *Lawrence and Travelogues*

Unlike Prichard, who in his *Researches* developed a sophisticated method for interpreting and comparing cultural phenomena such as language and mythology, Lawrence focused more extensively on the physical structure of humans in his lectures on the *Natural History of Man*. One reason for this difference in focus was because his work was originally part of a medical and natural history lecture series. Consequently, an analysis of linguistic evidence was largely underdeveloped in Lawrence's treatise. However, a close textual examination of primary sources was not absent from his study, and he formulated for readers an analytical technique that provided them with instructive questions to consider when interpreting ethnographic data. He wrote,

> What climates, what degrees of heat and cold, can man bear? How is he able to endure all the diversified external influences of such various abodes? Is he indebted for this privilege to strength and flexibility of his organization, or to his mental functions, his reason, and the arts which he has thence derived? Is he a species broadly and clearly distinguished from all others; or is he specifically allied to the orang-utang and other monkeys? What are his corporeal, what are his mental distinctions? Are the latter different in kind, or only superior in degree to those of higher animals? Is there one species of men only, or are there many distinct ones? What particulars of external form and inward structure characterize the several races? What relation is observed between the differences of structure, and those of moral feeling mental powers, capability, and actual progress in arts, sciences, literature, government?[133]

When reading the accounts of different extra-Europeans, researchers were to take note of their natural environment, social organization, material culture, cognitive ability and physical appearance.

It was not Lawrence's intention to get his readers to respond to each of his queries in the same manner that informants were to follow the directions of natural history questionnaires. Instead, these instructive questions were designed to teach his readers how to engage with the material and draw significant insights from it. Training in anatomy and physiology was of particular importance for this approach. He wrote,

> I mention [these questions] ... only as examples; and I take the liberty of adding my firm conviction, that these and similar matters will never be cleared up, except by those who are thoroughly acquainted with the anatomy and physiology of our frame; with comparative anatomy; with the principles of general physiology, and the analogies derivable from the whole extent of living nature.[134]

Accordingly, Lawrence believed that those lacking a background in anatomy and physiology would produce inferior results. Such a stance was almost certainly part of his vocational strategy. After all, when he initially wrote his lectures on the *Natural History of Man* during the 1810s, they were designed for medical students in London. It made tactical sense to position medically trained researchers at the forefront of the emerging research field, with himself as a leading figure.

Observation was a chief preoccupation of Lawrence's ethnological paradigm, and he recognized that because the study of human diversity was still in its infancy – with few practitioners – it was important to follow the example of skilled observers from other natural history fields. He wrote,

> The natural history of man is, indeed, yet in its infancy; so that a complete view of the subject could not be attempted. The description and arrangement of the various productions of the globe have occupied numerous observers in all ages of the world; and have engaged their attention so exclusively, that they have had no time to think of themselves. Every reptile, bird, insect, plant, even every mineral, has had its historian, and been described with minute accuracy; while the human subject has been comparatively neglected.[135]

Lawrence was particularly concerned that many of the large natural history compendiums produced in Britain contained little (if any) information on human varieties. For example, referring to George Shaw's work *General Zoology, or Systematic Natural History* (1800), he wrote,

> In a voluminous work now publishing in this country entitled General Zoology, or Systematic Natural History, man is altogether omitted, without notice or apology. Accurate, beautiful, and expensive engravings have been executed of most objects in natural history, of insects, birds, plants; splendid and costly publications have been devoted to small and apparently insignificant departments of this science; yet the different races of man have hardly in any instance been attentively investigated, described, or compared together ...[136]

During the first half of the nineteenth century, many naturalists were hesitant to include humans in their works, arguing that they held a privileged position within the natural world separate from the rest of the animal kingdom. Responding to these naturalists, Lawrence hoped that by raising awareness of this lack of coverage, more naturalists would contribute to the study of human variation and include it in their voluminous works.

Lawrence was also concerned about the quality of the observations Europeans recorded in their travelogues. As was the case with natural history, many travel accounts contained minimal descriptions of extra-European peoples. He noted that in cases where travelogues did examine the material culture of indigenous people, there was no discussion on the actual inhabitants that produced these artefacts. He stated, 'Even intelligent and scientific travellers have too often thrown away, on dress, arms, ornaments, utensils, buildings, landscapes, and obscure antiquities, the utmost luxury of engraving and embellishment; neglecting entirely the being, without reference to whom none of these objects possess either value or interest'.[137] Lawrence aimed to improve the calibre of travel accounts by further identifying ways in which voyagers could produce more concise and detailed descriptions of human varieties.[138] He hoped that 'the more accurate accounts of various people procured by modern travellers' would add to the 'progress of science and discovery'.[139]

There are several significant points to draw out when examining Prichard's and Lawrence's use of linguistics, philology, historical records and travel accounts. For instance, their comprehensive frameworks demonstrate the diversity of knowledge early ethnologists utilized when attempting to trace the origin of humans and classify them into groupings based on their physical conformation and language. No one field had sufficient evidence to support monogenetic or polygenetic arguments, but by using analytical models and data from a variety of research fields, it was possible to formulate a framework for studying races. However, because Prichard and Lawrence remained in England for the majority of their lives and never left the shores of Europe, they were reliant on all sorts of informants, including travellers, missionaries and colonial officers, for updated information on the races of the world. Thus early on it was evident that ethnology took full advantage of a complex colonial network for acquiring its data. Consequently, as we shall see in the following chapter, because of this dependence on informants living abroad, ethnologists developed a network for exchanging ethnographic data akin to what other early nineteenth-century natural sciences – such as botany – utilized.[140]

## Conclusion: The Emergence of the British Ethnological Community

On the evening of 22 December 1848, Prichard fell ill and died of pericarditis at his home in London at the age of sixty-two. It was a sudden and tragic end to the illustrious career of one of Britain's leading ethnological figures from the first half of the nineteenth century. It was Prichard and not Lawrence who became the face of British ethnology, and members of the community canonized him as the father of the nascent science. In 1850 the *Journal of the Ethnological Society of London* published an obituary in memory of Prichard, written by his fellow physician and ethnologist, Thomas Hodgkin. Saddened by the loss of his close friend, Hodgkin remarked that 1848 would forever be remembered by the scientific and medical world as the year 'the greatest writer who has treated the science of Ethnology, and investigated and classified the nations and kindreds and tongues of voice-varying men', died prematurely.[141]

Hodgkin continued by emphasizing Prichard's role in the development of British ethnology. He wrote that over the course of thirty years, Prichard 'had not only collected a great amount of valuable materials, but had brought to bear upon the difficult questions which his subject presents a variety of collateral knowledge for their elucidation, thereby not only enhancing the value of his own researches, but pointing out to future inquirers the path to truth, in which he made such important advances'.[142] With the death of Prichard, his followers at the Ethnological Society of London had the task of continuing to champion his monogenetic paradigm and his ethnological methodology. Figures such as Hodgkin, John Crawfurd (1783–1868) and Robert Gordon Latham represented the future leaders of the nascent discipline.

This chapter examined early observational techniques for studying human diversity in Britain during the first half of the nineteenth century. Natural history, medicine and travelogues all fed into early ethnological models, and each of these traditions shaped the study of races in different ways. The systematization of data was a key preoccupation during this period, and Prichard and Lawrence incorporated the classificatory practices of naturalists such as Linnaeus, Blumenbach and others into their ethnological frameworks. They were also proponents of monogenism and used natural history classificatory terms such as 'species' and 'varieties' in order to demonstrate that all humans shared one ancestral origin. Comparative anatomical studies were also a significant component of early ethnology. Prichard's and Lawrence's knowledge of anatomy was a major focus in their descriptions of human varieties. In particular, both figures further refined their observational technique by building upon Blumenbach's methodology as outlined in his treatise *De Generis Humani Varietate Nativa*. They also used the observations of voyagers from travelogues as natural history data to support their ethnological suppositions. Because neither scholar ever left the shores of Europe, they relied on the direct experience of explorers for evidence. In doing so, both Prichard and Lawrence believed their representation of human varieties were accurate and credible.

Between the late eighteenth century and the beginning of the Victorian period, the British global empire expanded. With it, a wealth of ethnographic material returned to Britain. Prichard and Lawrence were some of the earliest ethnological practitioners to incorporate this valuable resource into their studies. The result was two comprehensive and popular studies on human races, which constructed detailed taxonomical systems for interpreting the various peoples that inhabited the world, and connected these classificatory models to monogenetic frameworks in an effort to demonstrate the common origin of humans. Prichard published three editions of his *Researches* between 1813 and 1847, and there were nine unofficial reprinted editions of Lawrence's *Natural History of Man* published between 1819 and 1848.[143] Their success as writers and disseminators of ethnology in Britain is undeniable. Translated versions of Prichard's and Lawrence's works travelled to the European continent and North America, further extending their influences. Figures such as Prichard and Lawrence were not simply passive observers and anthologists of ethnographic material, but rather – as William Benjamin Carpenter put it – they were 'architects of Man' who actively engaged in developing ethnological theory and practice.[144] Both figures were highly attuned to the problems associated with establishing the observational accuracy of the emerging discipline, and they aimed to strengthen the criteria on which practitioners based their works.

# 2 ETHNOLOGY IN TRANSIT: INFORMANTS, QUESTIONNAIRES AND THE FORMATION OF THE ETHNOLOGICAL SOCIETY OF LONDON

> The Ethnological Society will supply the want; it will be a focus for travellers where, however scanty their contributions, a permanent value will be put upon their labours.[1]
>
> Richard King, 1844

In 1844 the ethnologist, surgeon and Arctic explorer Richard King (1811–76) delivered the first anniversary address before the members of the Ethnological Society of London (ESL). In his speech he argued that acquiring ethnographic materials was a primary concern for the discipline if it wanted to establish and refine its theories and methods. To meet this need, the ESL began to organize a global exchange network of informants – including colonial officers, military surgeons and missionaries – whose task was to collect ethnographic evidence *in situ* and send their reports back to Britain, where ethnologists could use it in their inquiries. Using travellers' accounts by voyagers was nothing new for ethnology, and since the expansion of the empire from the late eighteenth century onwards, practitioners interested in human diversity incorporated a wide range of travel material into their studies.[2]

However, early ethnologists such as William Lawrence lamented in his lectures on the *Natural History of Man* that the overall quality of many voyager accounts contained 'considerable discrepancies ... increased in many cases by haste and carelessness; by superficial examination, and loose choice of expressions'.[3] A chief problem was that in many instances informants gathering ethnographic data lacked formal training and had few instructions on what sorts of observations to record and what questions to ask. Consequently, in an effort to transform and refine the collecting practices of travellers, the British ethnological community began to produce a body of literature that taught informants how to observe and record a higher standard of ethnographic evidence for researchers back in Britain.

This chapter seeks to enlarge our understanding of early ethnological observational practices by demonstrating how British ethnologists in the 1840s continued to transform and refine the earlier methods of researchers such as Prichard and Lawrence. To this end, it will examine the influence of French ethnology on the British tradition. Members of the ESL recognized the importance of having reliable and consistent ethnographic observations for their studies, and they aimed to standardize the collecting practices of informants working abroad. Building on the work of historians Michael Bravo, James Urry and others, this chapter will show the central role the ESL played in this process.[4] In particular, it will examine the efforts of figures such as King, Thomas Hodgkin and John Crawfurd. Bravo remarked that 'The success of ethnological research was dependent on gaining reliable access to information about the people of other nations'.[5] By looking at the attempts of early practitioners to improve the quantity and quality of ethnographic observations returning to Britain, we gain a wider understanding of the way in which these researchers engaged with their object of study.

With these concerns in mind, this chapter will focus on the establishment of the ESL and the role ethnographic manuals and questionnaires played in this process. These instructional texts functioned as a statement of intent for British ethnologists. It demonstrated ethnology's dedication to raising the scientific criterion of its research by standardizing its methods for acquiring data. This was similar to how other natural sciences – such as botany – refined their collecting practices in the first half of the nineteenth century.[6] By standardizing the way in which informants recorded ethnographic observations, researchers strengthened their analyses of human varieties. If each informant collected data by following the same set of questions, ethnologists hoped that they in turn would be able to conduct more rigorous cross-comparative examinations of their materials.

## The Emergence of Ethnological Communities in Europe

> The main object, therefore, that should today occupy the attention and zeal of a truly philosophical traveller would be the careful gathering of all means that might assist him to penetrate the thought of peoples among whom he would be situated, and to account for the order of their actions and relationships.[7]
>
> Joseph Marie de Gérando, 1800

### *French Ethnology and the Société des Observateurs de l'Homme*

French ethnology had a significant influence on early British ethnological methods during the first half of the nineteenth century. British practitioners such as Hodgkin based their ethnological questionnaires on the works of French scholars and organized their learned bodies on the guiding principles of the earlier

French societies. The emergence of the French ethnological community developed fifty years before their British counterparts. George Stocking noted in his essay 'French Anthropology in 1800' that France had the world's first anthropological society.[8] There is a general tendency by scholars interested in the history of British anthropology to overlook the connection and influence of French ethnological and anthropological practice and theory on the British tradition.

The history of French anthropology differs significantly from that of its British counterpart. Most notably, in France there is a separation that exists between cultural theorists and ethnographers. Robert Parkin noted recently that 'there has always been a clearer division of labour in France than elsewhere between those who have produced major theories and methodologies of significance to anthropology and those who have collected ethnographic data in the field'.[9] This distinction between cultural theorist and ethnographer is still present within the French tradition today. As a result, the secondary literature on the history of French anthropology does not devalue the contributions of early ethnologists in the same way as their British counterparts. As an alternative, using a historicist approach, French scholars are more interested in examining the methods and theories of early nineteenth-century practitioners as important developers of the science.[10]

William Frédéric Edwards (1777–1842) founded the Société Ethnologique de Paris in 1839, and its members devoted themselves to the study of the various races that inhabited the world, examining their intellectual, physical and linguistic features. The Société Ethnologique de Paris was the product of a growing French interest in the study of human diversity. By 1822 there were already two learned societies interested in studying the cultural and physical characteristics of races. There was the Société Asiatique, which studied Asian art, literature and social structures in places such as India and China, and the Société Géographique, which focused on the various regions of the world that humans inhabited. French practitioners interested in human diversity belonged to a wide range of backgrounds and specializations. However, neither the Société Asiatique nor the Société Géographique focused specifically on ethnographic or ethnological research, and it was for this reason that the French community founded the Société Ethnologique de Paris in the 1830s.

However, the Société Ethnologique de Paris was not the first French society to specialize in studying human variation, and its guiding principles and practical framework came from an earlier learned community. Between 1799 and 1804 there was the short-lived Société des Observateurs de l'Homme, and it was at this society that a rigorous ethnographic methodology first developed.[11] French ethnologists had been producing instructive guides for researchers since the beginning of the nineteenth century.[12] According to Paul Rabinow in his book *French Modern*, historians of anthropology should look at the activities and efforts of the Société des Observateurs de l'Homme as 'a forerunner of modern, fieldwork-oriented ethnography'.[13] The active engagement of this society with both writing ethnographic manuals and sending informants to remote locations – such as the South Pacific – to collect ethnographic data on extra-European peoples demonstrates its dedication to establishing and refining the emerging discipline's theories and practices.[14] Among the members of the Société des Observateurs de l'Homme were some of the most eminent researchers on the subject of human diversity in France – including its founder, the literary figure and publisher Louis François Jauffret (1770–1850), the naturalist Jean Baptiste de Lamarck, the zoologist Étienne Geoffroy de Saint-Hilaire, the comparative anatomist Georges Cuvier and the botanist Antoine Laurent de Jussieu (1748–1836).[15]

In addition to being the first learned society strictly devoted to the study of human diversity, the Société des Observateurs de l'Homme produced two important texts for ethnographic research in 1800. The first was a short memorandum written by Cuvier entitled *Note Instructive sur les Recherches à Faire Relativement aux Différences Anatomiques des Diverses Races d'Hommes*, which focused on the anatomy and physiology of races.[16] The second was a guide written by the philosopher and jurist Joseph Marie de Gérando (1772–1842) entitled *Considération sur les Diverses Méthodes à Suivre dans l'Observation des Peuples Sauvage*, which provided informants travelling abroad with a detailed list of questions to consider when collecting data on human varieties.[17] The Société des Observateurs de l'Homme gave both Cuvier's and de Gérando's texts to Captain Nicolas Thomas Baudin (1754–1803) for his expedition to Australia, which set sail in 1800. Napoléon Bonaparte (1769–1821) financed the Baudin Expedition (1800–4) for the expressed purpose of collecting natural history data and charting the western coastline of Australia.[18] Baudin's crew included twenty-three naturalists, each instructed to collect specific types of data for different research fields, including astronomy, botany, entomology, geography, mineralogy, zoology and ethnology.[19]

The establishment of the Société des Observateurs de l'Homme at the beginning of the nineteenth century laid the organizational framework for future societies in both France and Britain to mirror. French ethnologists were committed to raising the overall quality of the observations recorded by informants travelling abroad. Not only did the Société des Observateurs de l'Homme establish a large network of informants around the world whose task was to collect data for French researchers in the metropole to utilize in their studies, the society also provided these voyagers with instructions on how to do it. Questionnaires such as de Gérando's text occupied a key role in this process. These instructive guides also provided later British ethnologists with a model upon which to base their own didactic literature.

## *British Ethnology and the Ethnological Society of London*

Before the ESL formed in 1843, many ethnologists in Britain discussed topics relating to human diversity at several societies. The philologist and orientalist Henry Colebrooke (1756–1837) founded the Royal Asiatic Society (RAS) in 1823, which was devoted to the study of Asian science, literature and art.[20] The RAS was not strictly interested in studying human variation in Asia, and its members discussed other natural sciences such as botany, geography and zoology. Another British society with links to ethnological studies was the Aborigines Protection Society (APS). The APS was dedicated to the preservation and protection of indigenous populations throughout the empire. Its motto was the Latin phrase *ab uno sanguine*, which translates into 'of one blood', relating the idea of a common human ancestral origin. Members of the APS were committed to the moral, physical and social improvements of extra-European peoples.[21] However, as Robert Kenny noted, 'the APS, with its philanthropic and political priorities, was limited as a scientific society'.[22]

Responding to these scientific limitations at the APS, Hodgkin and King organized a new group named the ESL that met weekly.[23] The purpose of this new society was to create a forum where 'gentlemen of science' could discuss the anatomy and physiology of different races, examine historical records, trace the form and meaning of various languages, present papers, and debate issues such as whether all humans shared a single ancestral origin.[24] Another central objective of the society was to produce ethnological texts to circulate throughout Britain and its empire. The ESL identified a connection between the circulation of knowledge through print and the expansion of their nascent science. Establishing informant networks and promoting the collection of ethnographic materials from the colonies were seen as priorities.

In the first volume of the *Journal of the Ethnological Society of London* (*JESL*) published in 1848, the ESL outlined its primary aims: 'The Ethnological Society of London is formed for the purpose of inquiring into the distinguishing characteristics, physical and moral, of the varieties of Mankind which inhabit, or have inhabited, the Earth; and to ascertain the causes of such characteristics'.[25] The executive council wanted to expand ethnological knowledge in Britain, and they recognized the important role a society strictly devoted to studying human diversity played in disseminating this information. From the outset the society recognized the value of earlier ethnological methods, and building upon the works of figures such as Prichard and Lawrence, its members used physical evidence such as anatomical and physiological descriptions of races and historical and linguistic evidence in their inquiries.

Large ethnological treatises such as Prichard's *Researches* and Lawrence's *Natural History of Man* incorporated a wide variety of evidence into their studies and utilized different theories and practices from several traditions, including natural history, medicine and travel literature.[26] This diversity of evidence was indicative of the state of ethnology during the early period. Prior to the publication of the *JESL*, ethnological literature was scattered across several genres, including geographical and philosophical journals.[27] Prichard, as an example, published articles in the *Annals of Philosophy*, the *Statistical Journal* and the *Journal of the Royal Geological Society*.[28] The publication of the *JESL* was significant because it created a specialized genre strictly devoted to expanding ethnographic and ethnological knowledge.[29]

In King's opening address delivered at the first anniversary meeting of the ESL, he stated, 'While other branches of Natural History are diligently cultivated, no degree of general attention, proportionate to its importance, is given to ETHNOLOGY, or the Natural History of Man'.[30] Recalling a prospectus he wrote the previous year, King discussed the original aims of the ESL. He began by arguing that 'among the numerous Literary and Scientific Societies established in the British Metropolis, one is still wanting to complete the circle of Scientific Institutions, whose sole object should be the promotion and diffusion of the most important and interesting branch of knowledge – ETHNOLOGY'.[31] King's prospectus corresponded with criticisms periodical reviewers levelled at large ethnological works during the first half of the nineteenth century. These reviewers routinely identified inconsistencies in the observations of informants who provided British researchers in the metropole with their data. For instance, after the completion of Prichard's third edition of the *Researches* – which he published as five volumes between 1836 and 1847 – William Benjamin Carpenter wrote for the *Edinburgh Review* that ethnology relied on the observations of untrained travellers. As a result, there were many incongruities in ethnological works.[32]

Dissatisfied with these reviews, figures such as King wanted to improve the quality of ethnological investigations by developing a society that was devoted to establishing and promoting the study of human races. Ethnologists hoped to build a community of trained researchers and informants who would collect ethnographic materials throughout the empire and either send their observations to ethnologists back in Britain to interpret and write ethnological accounts, or upon their return, present their ethnological observations at the ESL and produce their own written reports of their experiences with extra-Europeans.

King understood that for ethnology to be transformed into a scientific discipline, it had to refine the process by which it collected its data, analysed the material, and represented its findings. In many respects his ambition to improve ethnographic collecting methods was a statement of intent that British ethnology was dedicated to improving its observational practices. It was similar to the way other natural sciences such as botany developed its research techniques at the beginning of the nineteenth century, by getting collectors living throughout the world to acquire botanical specimens and send them to researchers back in Britain to utilize in their studies.[33] Darwin, for instance, relied on a large network of correspondents situated throughout the world for his data and specimens. He regularly sent requests to Europeans living in colonial settlements for their observations on different plants and animals.[34]

According to King, the first step in improving ethnology's research practices was to produce instructive texts on data collection and circulate them to a growing readership interested in human variation. Once the data was collected, he wanted the material to be stored in a centralized location so that its value could be maximized by researchers. King stated that 'there is a great amount of Ethnological Information existing in Great Britain, yet it is so scattered and dispersed, either in large books that are not generally accessible, or in the bureaux of the public departments, or in the possession of private individuals, as to be nearly unavailable to the public'.[35] Because the discipline incorporated data from several different research fields, including various natural sciences, medicine, geography and philology, it is difficult to identify any specific group as the primary candidates of King's research scheme. He did not refer to any specific type of researcher as being best suited for ethnological inquiries. However, we can deduce what sorts of occupations some ethnologists practised by looking at the society's membership and contributors. This is revealing because it demonstrates ethnology's strong links to natural history, medicine and imperial exploration.

King trained as an apothecary and as a surgeon. He gained further practical experience in ethnography as an Arctic explorer, while working as the surgeon and naturalist aboard George Back's expedition to the Northwest Passage in 1833.[36] King was also a member of the Royal College of Surgeons and the Society of Apothecaries.[37] Medical practitioners had a strong presence at the ESL, and in addition to King, the society's co-founder Hodgkin also trained in medicine at the University of Edinburgh, graduating with an MD in 1823. He went on to lecture on 'morbid anatomy' at Guy's Hospital in London in 1825, and by 1842 he was a teacher at St Thomas's Hospital in London.[38] Another leading member within the ESL was John Crawfurd, who trained in medicine at Edinburgh and travelled to Calcutta in 1803, where he became a surgeon in the British military. As was the case with King, Crawfurd gained considerable ethnographic experience from his travels through Asia, and he acquired an extensive knowledge of Malay languages. Upon his return to Britain, Crawfurd became a central figure at the ESL, regularly contributing papers to its periodical the *JESL*, and he was elected president in 1859.[39]

Looking at the table of contents from the first volume of the *JESL* in 1848 (see Figure 2.1), we can see more strong links to medicine. Hodgkin and Ernest Dieffenbach (both trained physicians) wrote the first two articles, 'The Study of Ethnology' and 'The Progress of Ethnology', which – as the titles suggest – discussed the state of ethnology in Britain during the 1840s.[40] Prichard also wrote an article on the significance of British ethnology for the natural sciences. In fact, the main focus of the first volume was to both define the research scope of the nascent discipline and promote its utility. There were other important contributors with backgrounds in medicine that submitted articles to the first volume of the *JESL*. The physician and philologist Robert Gordon Latham wrote two papers for the inaugural edition.[41] Latham was particularly important because he was a major figure to champion ethnology's monogenetic paradigm after Prichard's death in 1848.[42] Additionally, both Crawfurd and King contributed pieces.[43]

Ethnology's links to medicine were clear. With their educational backgrounds, physicians and surgeons were attracted to ethnology because it looked at the physical structure of different races. In addition, many ethnologists – such as King and Crawfurd – gained first-hand experience with extra-European peoples while working abroad as surgeons. Much of the evidence ethnologists used in their investigations came from the observations of informants with medical training who were working on naval ships that were encountering new populations inhabiting uncharted territories. For example, we saw in the last chapter that Prichard used the observations of the physician and naturalist Johann Reinhold Forster from Cook's second and third voyage to the South Pacific in his *Researches*.[44] Equally, Lawrence argued in his lectures on the *Natural History of Man* that medical practitioners were well suited for ethnological research because of their observational training in anatomy and physiology.[45] For these reasons, medical practitioners were well placed to collect ethnographic data and write ethnological accounts.

In King's prospectus – reprinted in his 1844 address – he stated four objectives for the ESL. Firstly, the society was 'To collect, register and digest, and to print for the use of Members and the public at large, in a cheap form, and at certain intervals, such new, interesting, and useful facts as the Society may have in its possession'.[46] Introducing ethnology to a larger audience was a central concern for the ESL because the study of human diversity required the assistance of informants living throughout the empire to furnish it with data for ethnological studies. The ESL wanted to continually attract new contributors. However, the executive council also understood that ethnology had to appeal to the interests of a non-medical audience. Not every colonial settlement had a surgeon or physician able to collect ethnographic data. It was therefore disadvantageous for the ESL to be overly selective and strictly rely on medical practitioners to acquire its evidence. Responding to this issue, ethnologists had to demonstrate to other potential informants – such as colonial officers – that topics relating to human variation possessed 'interesting, and useful facts' that were valuable for the advancement of the empire. Accordingly, the ESL could help government administrators better understand indigenous populations and assist in controlling and managing their settlements.[47]

# CONTENTS.

| | Page |
|---|---|
| Regulations of the Society, . . . . . | 3 |
| The Study of Ethnology. By Ernest Dieffenbach, M.D., | 15 |
| The Progress of Ethnology. By Thomas Hodgkin, M.D., | 27 |
| The Physical Characters of the Esquimaux. By Dr King, Honorary Secretary of the Ethnological Society, . . | 45 |
| The Yamúd and Goklan Tribes of Turkomania. By The Baron Clement Augustus de Bode, . | 60 |
| The Ancient Peruvians. By Dr J. J. de Tschudi, . | 79 |
| The Mongols. By Bayle St John, . | 86 |
| The Bilúchi Tribes inhabiting Sindh, in the lower Valley of the Indus and Cutchi. By Captain T. Postans, . | 103 |
| The Intellectual Character of the Esquimaux. By Dr King, Honorary Secretary of the Ethnological Society, . . | 127 |
| The Languages of the Oregon Territory. By R. G. Latham, M.D., | 154 |
| The Ancient Inhabitants of the Canary Islands. By Thomas Hodgkin, M.D., . . . . . | 167 |
| The Ethnography of Russian America. By R. G. Latham, M.D., | 182 |
| The Human Mouth. By Alexander Nasmyth, . . | 192 |
| The Natives of Old Callabar, West Coast of Africa. By W. F. Daniell, F.E.S., . . . . . | 210 |
| The Indian Tribes Inhabiting the North-West Coast of America. By John Scouler, M.D., F.L.S., . . . | 228 |
| The Natives of Guiana. By Sir Robert Schomburgk, . | 253 |
| The Industrial Arts of the Esquimaux. By Dr King, Honorary Secretary of the Ethnological Society, . . | 277 |
| The Relations of Ethnology to other Branches of Knowledge. By James Cowles Prichard, M.D., President of the Ethnological Society, . . . . . . | 301 |
| The Malayan and Poylnesian Languages and Races. By John Crawfurd, . . . . . . | 330 |

Figure 2.1: Table of contents from the first volume of the *JESL* in 1848. From Anon., 'Front Matters', *Journal of the Ethnological Society of London*, 1 (1848), pp. i–iii, on p. iii. Reproduced from the author's private collection.

The second objective of the ESL was 'To accumulate gradually a Library of the best Books on Ethnology – a Selection of the best Voyages and Travels – a complete Collection of Dictionaries and Grammars bearing upon the subject – as well as such documents and materials as may convey the best information to persons intending to visit Foreign Countries'.[48] Prichard and Lawrence used travelogues as ethnographic sources in their writings, and ethnologists at the ESL were building on these earlier methodological frameworks. The use of travel accounts authenticated ethnological research because many ethnologists had little first-hand experience living among extra-European peoples. It was equally important that the ESL possessed a comprehensive library so that its practitioners had access to the best available literature on the various races of the world. By having such a collection, its members could produce as accurate an ethnological study as possible.

The ESL strongly emphasized the significance of learning and documenting the languages of extra-Europeans because for monogenists it held the key to uniting the various races of the world to a single ancestral origin.[49] Once again the ESL was building upon earlier ethnological methods. The standardization of ethnological terminology was similarly important, and having consistent definitions of indigenous languages was essential for the research field. Lawrence explained in his lectures on the *Natural History of Man* that carelessness of language when describing races could lead to misrepresentations of indigenous populations.[50] Language was also a primary concern for French ethnology, and didactic literature from France influenced British practices in the early period. As we shall see shortly with de Gérando's questionnaire, knowledge of extra-European languages enabled the 'voyageur philosophe', who was collecting data, to better integrate into the indigenous communities and comprehend their cultural practices and symbolisms.

This emphasis on learning different languages was an attempt by early practitioners to help limit the amount of errors they produced in their cultural descriptions. It was a recognition that extra-European meaning systems varied extensively from European morals, values and customs. The third aim of the ESL's prospectus was most relevant to data collection, and King wrote that the ESL was 'To render pecuniary assistance, when the funds will permit, to such Travellers as may require it, in order to facilitate this particular branch of their research'.[51] Whenever it was possible, the ESL wanted to invest in the training of ethnographic researchers and informants.

In his last address at the ESL before his death in 1848, Prichard discussed the rapid expansion of ethnology as a science in Britain and linked this development to the abundance of new ethnographic accounts returning to the metropole from travellers and explorers alike. He stated that the 'general diffusion of knowledge on this subject, and of the beneficial results likely to arise from associations like our own, is to be traced in the works of maritime exploratators [*sic*], or of persons who have either been sent out by Government, or induced by motives of individual enterprise, to embark on voyages of discovery to distant countries'.[52] With more ethnographic material available, ethnologists could continue to transform and expand their knowledge of the various races inhabiting the world.

Because of the size of its empire, Britain was in a favourable position to create a large network with its colonies, making possible the collection of ethnographic data on a grand scale. King wrote, 'Great Britain is at the present moment in communication with the uncivilized nations of the world to a far larger extent than any other power on earth'.[53] The ESL wanted to capitalize on this network, and thus its fourth objective was 'To correspond with similar Societies that may be established in different parts of the world, with Foreign Individuals engaged in Ethnological Pursuits, and with the most intelligent British residents in the various remote Settlements of the Empire'.[54] Fundamentally, these four objectives demonstrate that British ethnology was actively engaged in developing ethnographic knowledge, and furthermore, in compensating for the limited first-hand experience many of its researchers had with the races they studied. King's prospectus also shows that ethnologists were continually trying to refine and improve their methodologies.

King ended his 1844 address by identifying the important relationship the ESL could build with the government, and he stated that 'To the Statesmen, then, this Society will lend a helping hand when studying the Colonial possessions of which he is the guardian'.[55] In the same year, King was also put in charge of establishing links with other learned societies in Britain, and the ESL 'left to the discretion of the secretary [Richard King] to exchange the Society's Journal for those published by other Societies'.[56] This was to attract new members to the ESL, compare ethnological research with other fields of inquiry, and promote the utility of ethnography and ethnology for other disciplines.

King was not alone in recognizing the significance of Britain's empire. In Ernest Dieffenbach's article 'The Study of Ethnology' – published in the first volume of the *JESL* in 1848 – he argued that Britain 'above all others, seems to be qualified to open a new era for this important science [of ethnology]'.[57] Dieffenbach (1811–55), who was a German ethnologist and physician living in London, asserted that Britain was in a favourable position to collect ethnographic data because its empire swept 'the easternmost boundaries of Asia, and the westernmost of America'.[58] He continued by stating that 'No nation, therefore, has such opportunity for investigating the history of races, – for no nation is it so essential to be well acquainted with these races'.[59] In addition to recognizing Britain's distinctive position as a leading nation for the collection of ethnographic data, Dieffenbach also outlined the role of the ESL in this scheme.

Firstly, the ESL was to act as an intermediary disseminator of the ethnographic material dispersed throughout the empire. Its chief function was to receive the data informants collected abroad and organize this information at a centralized location for researchers interested in human diversity to utilize at their leisure.[60] Dieffenbach wrote, 'It must be the object of the Ethnological Society, therefore, to obtain an exact knowledge of any nation or tribe, as they exist at the present moment. Ethnology begins with Ethnography – with an authentic description of the physical condition of each nation: and for this purpose it will be necessary to collect every thing that will throw light on this subject'.[61]

Secondly, once travellers had collected ethnographic data *in situ*, their task was to bring the evidence back to Britain, where ethnologists at the ESL could interpret the information and produce ethnological representations of indigenous peoples from their findings. Dieffenbach argued that 'It suffices here merely to indicate the fields of investigation with which the Ethnological Society should be occupied: it will be the especial duty of the Council to promote the publication of interesting matter contained in the books of Travellers of different nations, and the illustration of their publications with accurately-coloured portraits of typical specimens'.[62]

From its formation, the ESL was dedicated to improving the collecting practices of its informants and to expanding ethnological literature in Britain. Its members recognized their important role in establishing a global network with informants in order to maximize Britain's ethnographic imperial resources. It demonstrated that ethnologists were committed to transforming the quality of its research by refining and promoting its observational practices. Furthermore, it shows that ethnologists were not passive actors within this process. It was not the case that they were simply anthologizing ethnographic material; they were central in shaping and defining the methods of its researchers and informants.

## Instructive Literature for Informants

> ... the powers of observation must be trained, and habits of method in arranging the materials presented to the eye must be acquired before the student possesses the requisites for understanding what he contemplates.[63]
>
> Harriet Martineau, 1838

### *French Ethnographic Questionnaires*

Early ethnologists were committed to the development of their research field, and they were eager to improve and expand their knowledge of human diversity. From the onset of the discipline's formation, it was evident that ethnologists wanted to connect their research to the larger medical and scientific communities, and there was an initiative to develop techniques for acquiring a higher standard of ethnographic data. Members at the ESL such as Richard Cull – who was the secretary of the society – Hodgkin and Prichard designed ethnographic questionnaires to distribute to travellers sailing to the various colonial outposts. For instance, after delivering a paper at the British Association for the Advancement of Science (BAAS) in 1839 'On the Extinction of Some Varieties of the Human Race', Prichard co-produced a questionnaire with other members of the British ethnological community entitled *Queries Respecting the Human Race to be Addressed to Travellers and Others*. He also wrote a revised version of this instructive text for John Herschel's edited volume *A Manual of Scientific Enquiry* from 1849.[64] Early ethnologists recognized that a primary weakness inherent in their studies was the reliance on the observations of untrained informants. Responding to this problem, there was a strong initiative to produce questionnaires for travellers, which served to improve the quality and objectivity of their ethnographic reports.

The British community, however, modelled their didactic ethnographic literature on the earlier writings of French naturalists who had been producing questionnaires relating to human variation since 1800. Robert Parkin argued in his recent essay that

> If not in innovative fieldwork, then, France can certainly claim priority when it comes to ideas that have informed anthropological theory and practice. Indeed it can claim a clear first in producing one early modern writer who combined an interest in the customs, of other, exotic, and strange peoples with an attempt to systematize knowledge about human societies in a more or less objective manner ...[65]

French ethnology had a significant impact on the British community during the 1840s. Hodgkin, for instance, initially came across French ethnographic guidebooks during his visit to Paris in 1839. After reading these manuals, he sought to develop an English adaptation. However, his version was not widely circulated until 1851, when ESL collaborated with a sub-committee at the BAAS and invested £12 into the production of a manual designed by Hodgkin and Cull. This instructive text outlined key questions for informants to consider when collecting ethnographic material *in situ* on extra-Europeans. It was the second time the BAAS worked cooperatively with British ethnologists in order to produce an ethnographic questionnaire. The ESL noted that 'The object in publishing these questions is to induce Consuls, political and other residents and travellers, to obtain precise knowledge in reply to them, and send it to a centre [in London]'.[66] Originally produced as a pamphlet, Hodgkin and Cull based the structure of their questionnaire on the queries from French ethnographic guides issued by learned bodies such as the Société Ethnologique de Paris.[67]

As Desmond and Moore showed in their book *Darwin's Sacred Cause*, one of the initial reasons for the BAAS investing in the production of ethnographic literature was to compete with French ethnology, which possessed a more developed network of ethnographic collectors.[68] A close inspection reveals the strong influence French didactic literature had upon the structure and content of British ethnographic questionnaires. Figures such as Prichard in 1842, and more directly, Hodgkin and Cull in 1851, were indebted to their French counterparts. Hodgkin cites his visit to Société Ethnologique de Paris in 1839 as a central factor in shaping his decision to produce a comparable manual for British practitioners.[69] As such, it is illuminating to compare British ethnographic questionnaires with those of the French.

De Gérando's guidebook entitled *Considération sur les Diverses Méthodes à Suivre dans l'Observation des Peuples Sauvage* from 1800 was particularly significant for building a structured framework for acquiring ethnographic data. It gave detailed instructions for informants to follow, explaining what sorts of observations to document and specimens to collect. Unlike British manuals, which were designed for anyone travelling abroad, de Gérando wrote his guidebook for a highly specialized type of informant called a 'voyageur philosophe'.[70] De Gérando wanted to produce an ethnographic manual that systematically described human variation in terms of both its physical and cultural characteristics. However, because the Société des Observateurs de l'Homme originally distributed de Gérando's text with Cuvier's *Note Instructive sur les Recherches à Faire Relativement aux Différences Anatomiques des Diverses Races d'Hommes,* there was minimal attention to anatomical studies in the questionnaire.[71] Instead de Gérando wanted his 'voyageur philosophe' to document the physical strength and fitness of extra-Europeans.

De Gérando began his instructional guide by identifying eight faults voyagers routinely committed in their travel reports.[72] What is significant about his critique was how surprisingly reflexive it was in relation to the methods by which travellers collected, analysed and represented their ethnographic materials. It demonstrated that de Gérando comprehended the subtleties that existed between different human varieties and that he was committed to improving the way in which Europeans understood these differences. His framework provided both later ethnologists at the Société Ethnologique de Paris and at the ESL with a solid ethnographic model to build upon.

The first fault that de Gérando identified with the observations of explorers was the incompleteness of their accounts.[73] He argued that most ethnographic observations of various races were incomplete because in many instances the informants collecting this data spent an insufficient amount of time engaging with their subjects. The second problem, according to de Gérando, was that informants were too hasty in ascribing characteristics to different indigenous peoples. They often made bold generalizations either based on European understandings of cultural values and customs, or by basing their interpretations on atypical individuals from the observed society. De Gérando wrote,

> such insufficient observations have not always been very certain or authentic, whether because they have sometimes been too particular, and explorers have wished to judge a society by a few of its members, a character by a few actions, or because they have confined themselves to hearsay, to the stories of the Savages ...[74]

It is surprising to see this level of reflexivity in de Gérando's work because it undermined the authority of the observations recorded by European explorers who were directly engaging with indigenous peoples. This is further significant because many ethnologists in the first half of the nineteenth century substantiated their research by using travel reports. However, de Gérando was trying to refine the observational practices of the research field, and thus he was not attempting to discredit these early studies on human diversity per se, but rather to improve on them.

The third and fourth faults, according to de Gérando, were that many informants failed to utilize a structured methodological framework for recording their information. Consequently, he aimed to establish a standardized approach for observing indigenous peoples, which would transform the traditionally unstructured and disjointed accounts of the explorers into the reflexive, organized and comprehensive observations of an informed 'voyageur philosophe'. De Gérando argued,

> We should add that these observations have been badly ordered, and even in many cases quite without order. The explorers had not enough understood that there is a natural connection between the various facts that one gathers about the condition and character of societies, that this order is necessary to the precision of the individual facts, and that often some of them should serve as preparation for the others.[75]

Furthermore, he stated that 'Often explorers have based the accounts that they bring us on incorrect or at least on dubious hypotheses. For example, they habitually judge the customs of Savages by analogies drawn from our own customs'.[76] Subsequently, de Gérando's instructive manual would help to minimize these shortcomings because the 'voyageur philosophe' was to follow the sequence of questions in their order, and in doing so, maximize the quality of their ethnographic descriptions. They would attempt to perceive the world according to the cultural explanatory model of the people under study. Thus, for example, a 'voyageur philosophe' would ask questions such as: how do the Maori interpret the ocean?

In order to accomplish such a task, and understand the way in which extra-Europeans understood the natural world, de Gérando wrote that the 'voyageur philosophe' had to learn the language of his subjects. This was discussed in de Gérando's fifth, sixth and seventh faults with the observations of explorers. Moreover, de Gérando argued that the choice of expressions used by travellers to describe certain rituals or beliefs often misinterpreted the significance of certain cultural acts. He wrote,

> In the case of the accounts of explorers there is another cause of uncertainty, a fault of language rather than of imperfect observation, namely that the terms used to pass on to us the results of their observations are often in our own language of vague and ill-determined meaning. Consequently, we are in danger of taking their accounts in a way which they did not intend. This happens particularly when they try to record the religious, moral and political beliefs of a people.[77]

Responding to these cultural misinterpretations, de Gérando argued that 'it is by learning their language that we shall become their fellow citizens'.[78] In doing so, the 'voyageur philosophe' would have a more comprehensive understanding of their cultural practices and symbolisms.

In terms of theory, de Gérando was not applying monogenism to his examination of extra-European languages. Cross-comparative linguistic and philological studies became a central tenet of British ethnology from the 1810s onwards.[79] Through language, ethnologists such Prichard argued that similarities between different lexicons demonstrated a common origin. For de Gérando it differed: language was a methodological tool that would serve to enhance the 'voyageur philosophe's' racial descriptions because he could avoid miscommunication between him and his subject by speaking in their local dialect. De Gérando wrote,

> But of all the regrets left by the accounts of the explorers who went before you, the strongest is their failure to tell us the language of the peoples visited. In the first place, the scanty information which they do give lacks precision and exactness, whether because they fail to record how they went about questioning the Savages, or because they themselves have often taken little care to pose the questions properly. The demonstrative and natural gestures which they have used to ask the Savages the names of objects were often themselves liable to considerable uncertainty ...[80]

In many respects, de Gérando's questionnaire was a fairly repetitive text. His queries began by instructing the 'voyageur philosophe' to ask questions about the language of the people under study. This included questions such as, 'Are the names of actions, or *verbs*, conjugated according to *person*, *tense*, and *mood*, or do they simply remain in the absolute?' He continued by writing, 'Are there *articles* and *conjunctions*, and of what Kind?'[81] This information was also useful for those researchers such as de Gérando and Cuvier who wrote about human diversity back in France. It was essential, according to de Gérando, that the 'voyageur philosophe' followed the order of questions so as to produce the highest quality of observations. De Gérando argued, 'We shall therefore confine ourselves to giving them a tabulated summary of the most important points at which their observations should be directed, to presenting them with the order which seems to us most appropriate, and to reminding them that the results reported to us will be valuable and useful in proportion to their degree of completeness'.[82] Once the 'voyageur philosophe' gained a competent understanding of the grammatical and syntactical rules of a language, and had a working knowledge of it in order to ask questions, his next task was to shift his focus to more culturally specific questions. De Gérando divided these queries into two themes, 'the state of the individual, and that of the society'.[83] In each case he discussed the individual and society in relation to both a physical and a moral existence.

The first two points that de Gérando discussed in relation to the physical existence of extra-Europeans related to climate and food. By studying the climate in which indigenous populations inhabited, the 'voyageur philosophe' could document the effects of weather conditions on the daily lives of the natives. This was also a chief component of Lawrence's ethnological framework from his lectures on the *Natural History of Man*. Lawrence viewed climate as a primary influence on the physical appearance of different human varieties.[84] In the case of food, de Gérando wrote that it was important to document 'the quality and quantity of his ordinary food and drink. A careful analysis of the water with which he quenches his thirst, and a test of his distaste for our own food'.[85] Both of these lines of questioning were also useful for colonial officials because these documented the local environmental conditions and natural resources of the land they occupied. In addition, it would potentially induce military and government officials to contribute further to investigations into the study of human races. It is significant that Napoléon Bonaparte financed the Baudin Expedition for the expressed purpose of collecting natural history data, and de Gérando had an obligation to satisfy the imperial ambitions of the French Empire.

De Gérando's questionnaire also instructed the 'voyageur philosophe' to document the physical existence of extra-Europeans. For anatomical and physiological instructions, his observer was to use Cuvier's didactic pamphlet. In de Gérando's text it differed: instead of asking the 'voyageur philosophe' to record the size and weight of the indigenous people, he wanted descriptions of the physical strength of indigenous populations. This was also illuminating for colonial officials because it showed the potential of using the local population as a cheap workforce for setting up European settlements. De Gérando wrote, 'We shall be given more positive information about the physical strength of the individual savage. You will discover what burdens he is capable of lifting, carrying, or dragging; what are his most successful muscular movements; how quickly he can run; how far he can travel without rest; how good he is at swimming; what physical exercises he ordinarily undertakes; you will observe how he climbs trees, crosses ditches, climbs rocks, and so on'.[86]

Clothing was also a topic discussed in de Gérando's questionnaire, and he stated that in many instances this was the most developed area of ethnographic observations because most accounts of extra-European peoples focused on their physical appearance. De Gérando argued that 'The clothes of Savages are generally very well described by explorers; this is nearly always their main observation, often the only one'. However, de Gérando wanted a more sophisticated description of indigenous clothing, and he stated that 'one should not confine oneself to observing their costume; you should investigate how repugnant they would find it to change their customs or adopt ours; and whether they have any reason for wearing what they do ... you should find out whether clothes vary with age or status, whether they give evidence of authority or wealth, and if there is any idea of adornment, and if so how important it is'.[87]

De Gérando further requested detailed observations about kinship relations within extra-European societies. He wanted the 'voyageur philosophe' to record fraternal relationships and the role of women within different populations. He wrote, 'What is the force and character of the bond between brothers? Is there any precedence of age among them? To what point do the relations of kinship extend and keep any influence?'[88] As was the case with his section on clothing, de Gérando wanted an analytical description of the role of women. Thus he instructed his 'voyageur philosophe' to document several features of women in indigenous societies. He wrote,

> The domestic condition of women can be considered from several points of view. The first to present itself is that of their dependence, or of the consideration shown for them. In this respect, Savages have sometimes presented quite different spectacles. In general, however, it seems that consideration for the female sex is an effect of civilization. The observations of explorers will teach us how far this idea has any foundation; they will teach us what right a woman has on common property, what work she has to do, what protection she is given; finally, they will teach us whether even in the most savage countries, the female sex preserves something of that sweet and secret power, rooted at once on her weakness, on her sensitivity and on her charms.[89]

Finally, de Gérando wanted to know how indigenous populations responded to, and interacted with, Europeans. In other words, was the observed community aggressive or hospitable? This was a chief concern for European empires who wanted to settle colonies in distant lands. A supportive relationship with native populations would help to stabilize and build colonial European communities abroad. There was a lot of tension between Europeans and indigenous populations during the late eighteenth and early nineteenth century, and many explorers had died in conflicts over cultural miscommunications with aboriginals. Cook, for instance, was killed in Hawaii in 1779 after the indigenous community mistook him for an evil deity.[90] Thus prior knowledge about extra-European treatment of explorers was useful for the empire. De Gérando wrote,

> What welcome does a savage people give to strangers quite unknown to it, like Europeans? What is the reason for this reception? Here careful precautions must be taken before making a judgement. One must try first to ascertain that the reception given one by a savage people is not the effect of memories left by other strangers; and then it would be necessary to know also what the actual conduct of these strangers

> towards the people was. Even when no memory could influence the reception that one receives, are there several ways of explaining the same reception? Fear and ferocity equally can put arms in the hand of the people visited; good nature, trust, timidity or perfidy can equally lead him to accord strangers a favourable welcome. Finally ferocity, defiance, or the sweet virtue of hospitality, are themselves moral phenomena whose causes we should, as far as possible, seek to penetrate.[91]

Looking at de Gérando's questionnaire is instructive because it shows the complexity of early ethnographic manuals in France. A close examination of these texts shows that practitioners interested in human diversity wanted to improve the quality of the ethnographic data informants collected abroad.

A chief difference, however, in the French approach to studying human variation was that the 'voyageur philosophe' was not simply collecting data for other researchers back in the metropole to incorporate into their studies, although figures such as Cuvier did use this material. By contrast, the 'voyageur philosophe' analysed the evidence and produced their own reports based on their findings. Thus one figure did all three components of the observational practice. Although Hodgkin was familiar with the ways in which French ethnologists conducted ethnographic surveys from his visit to Paris in 1839, his questionnaire – published in collaboration with the BAAS – differed from de Gérando's text. The British questionnaire was less reflexive about the ways travellers collected their observations. There was more flexibility for the British informant, and so they did not require the same level of training and specialization as the 'voyageur philosophe'. This was potentially done in order to encourage more people to use the text, and not just trained informants. Moreover, unlike their French counterparts who had government sponsorship for travellers and collectors, the British relied on volunteerism for acquiring data abroad. Anyone, regardless of their educational background, could record ethnographic data for British ethnologists. Perhaps figures such as Hodgkin and Prichard worked under the assumption that the observations of informants were objective descriptions and that once skilled researchers back in Britain analysed the material, they would be able to eliminate any incongruities within the reports.

## *Early British Ethnological Manuals*

One of the earliest ethnographic manuals produced in Britain was by the journalist, abolitionist and philosopher Harriet Martineau (1802–76). After a lengthy visit to the United States in 1834, Martineau took an interest in extra-Europeans. She noticed that in many instances the descriptions of travellers and explorers misrepresented indigenous populations. Thus in 1838 she published a manual entitled *How to Observe: Morals and Manners*. Its aim was to refine the observational practices of travellers who were collecting ethnographic material abroad. Martineau wrote,

> The observer of Men and Manners stands as much in need of intellectual preparation as any other student. This is not indeed, generally supposed, and a multitude of travellers act as if it were not true. Of the large number of tourists who annually sail from our ports there is probably not one who would dream of pretending to make observations on any subject of physical inquiry, of which he did not understand even the principles.[92]

Martineau argued that one of the best ways to develop ethnographic observational techniques was by refining one's methods prior to visiting colonial settlements. She believed that travellers should first visit locations in Britain and examine the local populations in order to develop the quality of their reporting. It was equally important to create a sense of reflexivity in the observational practices of aspiring ethnographers. She wrote,

> Which of us would undertake to classify the morals and manners of any hamlet in England, after spending the summer in it? What sensible man seriously generalizes upon the manners of a street, even though it be Houndsditch or Crandbourn-Alley? Who pretends to explain all the proceedings of his next-door neighbour? Who is able to account for all that is said and done by the dweller in the same house ... If such judgements were attempted, would they not be as various as those who make them? And would they not, after all, if closely looked into, reveal more of the mind of the observer than of the observed?[93]

Martineau further noted that the populations that informants described were only representing a sample of the community and did not necessarily exemplify the whole of a people. She stated that '[The observer] ... is furnished, at best, with no more than a sample of the people; and whether they be indeed a sample, must remain a conjecture which he has no means of verifying'.[94] Similar to de Gérando's questionnaire, Martineau's manual aimed to refine the observational practices of informants, and it reflexively considered the limitations and biases of European travellers.

Prichard also produced two ethnographic manuals for informants. The first manual was developed after he presented his paper 'On the Extinction of Some Varieties of the Human Race' at the BAAS meeting in 1839. A committee led by Prichard was set up to write a handbook that instructed voyagers on how to collect ethnographic data.[95] The guide was entitled *Queries Respecting the Human Race to be Addressed to Travellers and Others*, and the BAAS reproduced a copy in its annual report from 1842.[96] A more comprehensive version of the guidebook was produced in 1849, when the British mathematician, astronomer and photographer John Herschel (1792–1871) edited a book for the Royal Navy entitled *A Manual of Scientific Enquiry*. The aim of the manual was to standardize scientific data collecting by providing British officers and travellers with an instructive guide on how to accumulate evidence for various scientific fields. Prichard wrote the questions for the chapter on ethnology, and it was his final major contribution to the discipline before his sudden death in 1848.[97]

The manual was beneficial for several reasons. Firstly, the navy provided copies of Herschel's book for officers as they departed for the colonial world. This was significant because in principle it would maximize imperial resources by utilizing as many available informants as possible scattered throughout the empire. Secondly, it would transform the quality of the data returning to Britain because although the collectors for the most part were untrained officers, travellers and missionaries, the manual provided these informants with detailed instructions on what to look for and describe when examining indigenous peoples. Thirdly, because ethnology was in its formative years, it was impossible to make a living as a full-time practitioner. Many ethnologists were unable to travel abroad to remote locations and conduct their own observation *in situ*. Thus they were reliant on informant networks for information. Fourthly, it was a statement of intent that ethnologists were committed to improving the quality of the ethnographic data and recognized its shortcomings. Manuals such as Herschel's edited volume were important scientific tools because they afforded ethnologists with an opportunity to engage more directly with their data collectors. It also linked ethnology to other field sciences from the early nineteenth century – such as botany – which used imperial networks of exchange in order to expand botanical collections for researchers back in Britain. Ethnologists were building on existing practical models.[98]

The *Manual of Scientific Enquiry* appeared in three editions between 1849 and 1859, and Prichard's section on ethnology remained largely unchanged. However, because he wrote his chapter prior to his death in 1848, Thomas Wright – the secretary of the ESL – edited the section to ensure that its focus and methodology were up to date. The chapter was divided into twenty-six subsections and examined both physical and cultural features in different races. The first subsection, 'Objects of the Science of Ethnology', outlined the scope of ethnological research and its primary aim, which was to document the peculiarities of human diversity. Prichard stated that '[Ethnology] ... includes the physical history of man; that is, an account of the peculiarities of his bodily form and constitution, as they are displayed in different tribes, and under different circumstances of climate, local situation, clothing, nutrition, and under the various conditions which are supposed to occasion diversities of organic development'.[99] Prichard continued by writing that 'The second division of this general subject ... [is to study] ... the history of man as a social being'.[100] He organized his chapter by devoting eight subsections to topics relating to the physical characteristics of humans and fifteen subsections to cultural characteristics. The final two subsections discussed ethnological method.

In the subsection on the physical characteristics of different races, Prichard looked at 'Physical Characters of Nations', 'Complexion and Colour', 'Shape of the Features', 'Shape and Relative Size of the Head', 'Form and Structure of the Body' and 'Muscular Strength'. Once these observations were recorded, the next step was to examine the social organization of the people, including their language and material culture. Prichard outlined the types of cultural characteristics informants were to document, such as the 'Characteristics of the State of Society', 'Practice of Agriculture', 'Mechanical Arts', 'Metallurgy', 'Arts of War', 'Clothing', 'Domesticated Animals', 'Art of Navigation', 'Notions on Physical Science', 'Notions of Morality', 'Regulations of Social Life', 'Polygamy', 'Civil Government', 'Religious Impressions and Superstitions' and 'Language, Poetry and Literature'.[101]

Prichard wanted more comprehensive descriptions of extra-Europeans than the types of reports explorers such as Cook and his agents recorded. Ethnologists did not simply want reports of cross-cultural encounters with indigenous populations, with basic depictions of their physical appearances and social organizations. Rather, researchers interested in human variation wanted thorough documentation of everything, from the hair, eye and skin colour of the community to accounts of their religious practices. This information was not only valuable to ethnologists back in England; it was also useful for the British military, as it clearly outlined important information such as the technological prowess of native populations and their social organization. This connection to imperial ambitions is further strengthened by the fact that copies of the *Manual of Scientific Enquiry* were distributed to officers on board British naval ships. The final two subsections of Prichard's chapter explained methodological approaches for collecting the information. They were entitled the 'Importance of Obtaining Specimens of Composition' and 'Methods of Obtaining Information Concerning Languages in which no Written Compositions Exist'.[102]

Prichard's contribution to the *Manual of Scientific Enquiry* was significant for several reasons. Since ethnology was reliant on the efforts of travellers, explorers and missionaries alike to collect ethnographic material, producing a detailed guide outlining the sorts of evidence informants should collect would enhance the quality of ethnographic data returning to Britain. It also demonstrated an initiative to refine the emerging field of study. Ethnologists such as Prichard were proactive participants in producing didactic literature on ethnographic data collecting and not simply reactive observers and anthologists of the material. This suggests that historians should begin to recast early ethnology as more than simply an armchair pursuit. It was, after all, a discipline that prioritized *in situ* observation and utilized a comprehensive network of correspondents to acquire its data. With these disciplinary concerns in mind, ethnology as a science had much in common with other natural history fields such as botany and zoology. Practitioners interested in flora and fauna also relied on a complex network of correspondence for their evidence.

In 1851 the ESL and the BAAS commissioned Hodgkin and Cull to write an English guidebook for British ethnologists that was based on the French model. Hodgkin and Cull wrote that 'The Sub-Committee endeavoured, by various modifications, to render the new edition [of the Société Ethnologique de Paris's pamphlet] still more useful as a guide for studying man ethnologically. The title was changed, an edition of three thousand was printed for the £12, and is in course of circulation for the guidance of travellers, missionaries, and others, in collecting the kind of knowledge which we seek'.[103] Once information was collected, informants were to send their data back to the BAAS for ethnologists to interpret and produce ethnological accounts.[104] Copies were available through Hodgkin and John Arthur Phillips (1822–87), geologist and Fellow of the Royal Society.[105] Because the manual was produced collaboratively by the ESL and the BAAS, it was available at both societies' meetings. It is suggestive that Phillips had copies of the manual, and it indicates that geologists were seen as potential contributors to ethnology. During the nineteenth century, many naturalists contributed to the discussions within both of these research fields. Darwin, for example, was actively involved in ethnological and geological circles.

In 1854 the ESL reprinted Hodgkin and Cull's manual in the third volume of the *JESL*. There were eleven categories and one hundred and three queries for informants to consider when observing and collecting ethnographic data abroad. The instructions directed travellers to document any cultural or anatomical variance observed during their encounters with extra-Europeans. The guide stated that 'Any amount of knowledge, however trifling it may appear in itself, may be of great value in connexion with other knowledge, and therefore will be welcomed. We are seeking facts, and not inferences; what is observed, and not what is thought'.[106] Because many informants lacked formal training in medicine and natural history, the ESL did not want them to analyse the data they collected. That was a job for the skilled ethnologists back in the metropole. The primary aim of the informant was to collect empirically objective evidence. The BAAS and ESL were building a science, and in order to strengthen their analytical claims, the primary types of data sought were physical descriptions of races, documentation of languages, and observations of extra-European social organizations. Collecting the data *in situ* was only the first step in producing ethnological accounts. Informants did the groundwork, but it was up to the skilled ethnologist to interpret the material.

Unlike de Gérando, who emphasized language as the starting point for ethnographic observation, the first category in the British manual focused on the 'Physical Characters' of races. It contained sixteen points for informants to consider, and their first task was to 'Ascertain the form, size, and weight of [the] people' under study.[107] This was nothing new for British ethnology, and Prichard and Lawrence had discussed at length techniques for observing the physical conformations of human varieties.[108] Hodgkin and Cull also recognized the value of other types of ethnographic data in their manual, and they argued that 'The Ethnologist describes man as he is found in various localities, with all the peculiarities of physical structure and mental constitution'.[109] Consequently, the non-physical pronouncements of different races were also an important aspect of their ethnographic paradigm.

The second category from the manual focused on 'Languages', and it contained seven points. Hodgkin and Cull wrote that 'The affinity of languages is one line of evidence of high value in ethnological researches, and hence the importance of obtaining accurate information concerning the languages of a people'.[110] Again, this was not a new direction for ethnological research. De Gérando put a strong emphasis on language as a means for better understanding extra-European cultural practices and meaning systems. Prichard also regarded language as a central mode of investigation in his framework, and through linguistic studies he attempted to link all humans to a single origin. In 1831 he published a book on *The Eastern Origin of Celtic Nations Proved by a Comparison to Sanskrit, Greek, Latin and Teutonic Languages*. Its purpose was to show how similarities in languages demonstrated the unity of races.[111] Hodgkin and Cull were building on pre-existing French and British models. The third category of their manual discussed four points concerning grammar.[112]

Category four looked at 'Individual and Family Life', and it contained thirty-two points on kinship relations in different societies.[113] The informant was to look at relationships between parents and children, wives and husbands, and communal associations. Some of the questions considered were 'What is the ordinary size of families, and are there any large ones?' and 'Is polygamy permitted and practised, and to what extent?'[114] Studying social organizations was an important component of ethnological inquiries, and from it they could draw comparisons between different societies. Similarities in the division of labour between two neighbouring groups potentially demonstrated common ancestry. This information was also useful for the military because it provided officers and administrators with detailed descriptions of the indigenous populations that inhabited their territories.[115]

Category five, on 'Buildings and Monuments', contained three points for informants to examine, and category six looked at 'Works of Art' and had five points of consideration.[116] Understanding the material culture of extra-European societies was illuminating for ethnologists back in Britain because it potentially helped them interpret the values, traditions and histories of indigenous peoples. It also showed their technological prowess. For instance, point sixty-one requested information on the buildings indigenous groups inhabited.[117] Hodgkin and Cull also instructed their informants to look at the types of domesticated animals these societies possessed. Once again, this information was valuable for colonial administrators. Finally, the last four categories, 'Government and Laws', 'Geography and Statistics', 'Social Organization' and 'Religion, Superstition &c.', all considered different types of social institutions within extra-European cultures.[118]

Hodgkin and Cull's manual was comprehensive, and it aimed to enhance the overall quality of observations collected by travellers while living among indigenous communities. It was a clear attempt to standardize ethnographic practice because every informant would look for and collect similar types of evidence. This would in turn transform ethnological research back in Britain and enable them to conduct more rigorous cross-cultural comparative analyses of different races. However, unlike de Gérando's questionnaire, which began by reflexively considering the problems inherent in ethnographic collecting methods, Hodgkin and Cull did not consider how the ethnocentrism of travellers affected their observations. They did not warn their informants about European cultural biases. Perhaps this was because ethnologists such as Hodgkin presupposed that their analyses of the material would weed out their informants' misinterpretations.

It is difficult to determine the overall success of instructional manuals such as Prichard's 1849 chapter on ethnology from John Herschel's *Manual of Scientific Enquiry* and Hodgkin and Cull's guidebook from 1851. However, ethnologists and anthropologists recognized the utility of these texts and continued to produce guidebooks and manuals throughout the nineteenth century. The most famous of these books was *Notes and Queries in Anthropology*, first published in 1874 and reworked into five more editions between 1892 and 1951.[119] Evidence of the value of this didactic literature was already apparent in the 1880s when E. H. Man structured his research 'On the Aboriginal Inhabitants of the Andaman Islands' (1882) around the categories of *Notes and Queries*. It was an impressive description of the cultural and physical characteristics of the Andaman islanders.[120] Nevertheless, in the case of the manuals and guidebooks from the 1840s, we should think of them as guiding principles insofar as they were exceedingly ambitious. Even if the informant collecting the data only focused on specific sections of questionnaires, the quality of those observations would be seen as an improvement on the reports of travellers who did not have instructional guides.

## Conclusion: The New Voices of British Ethnology

During the 1840s ethnology in Britain expanded. Although Prichard and Lawrence were some of the most prolific British writers to tackle the subject of human variation in the earliest period, they were not alone. There was a growing community of scholars contributing to the study of races. In this chapter I examined the development of an organized British ethnological community during the 1840s. The ESL was the first British society strictly devoted to the study of human diversity, and they based their societal organization on the earlier established French model. The nature of the ESL illustrates ethnology's close links to medicine and imperial expansion. The ESL, for example, attempted to establish a global network of informants whose task was to supply ethnologists in Britain with data for their studies.

In December 1848 Prichard suddenly became ill and passed away. At the time of Prichard's death he was president of the ESL. Thus not only did the British ethnological community lose a leading figure who defined a prolific methodological framework for studying human diversity, they also lost their society leader. Consequently, the executive members asked Vice-Admiral Charles Malcolm (1782–1851) to take over as interim president of the ESL until the Fellows selected a new long-term replacement. Malcolm is a fairly obscure figure in the historiography. However, in the 1840s he was a central member of the ethnological community. Malcolm was president of the ESL prior to Prichard's election. He was an experienced traveller with a familiarity of the Indian Ocean region and served as the superintendent of the 'Indian Navy' in 1828.[121] With his knowledge of extra-European peoples, he was a suitable leader for the society, although he was in the twilight of his career.

Malcolm acknowledged that a central concern for the British ethnological community was that it required a new figurehead to carry the discipline forward. Now that one of the main voices of ethnology was gone, the community needed another ethnologist to become a guiding voice for the emerging science. Malcolm stated that 'As yet no one has come forward to fill [Prichard's] ... place. I was requested to be your President again. I accepted the honourable office with reluctance, and with an understanding that I was to retire as soon as you could find some eminent Ethnologist to take the Presidential Chair'.[122] It was clear that the ESL had to find a new president, but the problem was that there was a shortage of candidates wanting to fill the void left by Prichard. One person that Malcolm suggested was Alexander Von Humboldt (1769–1859), arguing that his knowledge of geography, philosophy, philology, physiology and history was almost unrivalled. However, as was the case with Malcolm, Humboldt was quite elderly by the early 1850s and thus not the best long-term solution.[123]

Eventually, the ESL elected Benjamin Collins Brodie (1783–1862) as president. Brodie appeared to be a sound solution for the ESL. As was the case with many of the early ethnologists, he had a background in anatomy and physiology. He trained as a surgeon at St George's Hospital in 1803 and became a Fellow of the Royal Society in 1810 and a Fellow of the Royal College of Surgeons in 1816. He was also appointed as an assistant surgeon at St George's Hospital in 1808 and was promoted to senior surgeon by 1822. Moreover, Brodie was an experienced speaker, having delivered surgical lectures at St George's Hospital from 1808 to 1830.[124] However, it was clear that he had larger ambitions, and his scholarly focus remained firmly in medicine and science generally. For example, by 1858 he became the president of the Royal Society. He did not become the new figurehead for British ethnology. Instead it was a group of ethnologists, including Hodgkin, King, Crawfurd and others from the executive council, who collectively contributed to the direction of the research field in the years leading up the anthropological schism of the 1860s.

However, one practitioner continued to produce large ethnological compendiums during the 1850s comparable with Prichard's work: a philologist and physician by the name of Robert Gordon Latham. The sciences of Man also faced a re-emerging adversary in the 1850s, which was the growing community of ethnologists interested in morphology and polygenetic theory regarding the origin of humans. One of the most significant figures within this group was the anatomist and physician Robert Knox, who published a controversial book on the study of humans entitled *Races of Man* (1850).[125] This text had a significant effect on the British ethnological community, and there was a growing animosity between ethnologists in the 1850s over how to conduct their studies.

# 3 ETHNOLOGY AT HOME: ROBERT GORDON LATHAM, ROBERT KNOX AND COMPETING OBSERVATIONAL PRACTICES

> Many well-educated persons, and some of them scientific men, have vague notions of our inquiry, and of the means of our investigation.[1]
>
> Richard Cull, 1851

In the summer of 1851, the secretary of the ESL, Richard Cull, presented an essay on ethnological practice and theory. Afterwards, he published his essay in the 1854 volume of the *JESL*. The main focus of the article was to discuss three continual challenges within the nascent research field. The first was to establish some kind of consensus on the types of evidence researchers were to incorporate into their studies. The second was to formulate an interpretive framework that assisted researchers with their analyses of data. Finally, the third was to identify a clear aim for using this material, including published works and exhibits showcasing foreign people. Cull's remark about the vague conceptual notion of what ethnology entailed is revealing because it brings to the fore a chief exegetic problem for early ethnologists. What was the best way to practise the science? With the growth of the British Empire, an abundance of new ethnographic material returned to Britain. Consequently, some members of the British ethnological community wanted to collect as much of this ethnographic data as possible and store it at a centralized location within the ESL. In doing so, its members and other interested parties would have access to a comprehensive collection in order to further refine and substantiate ethnological discourse and knowledge.

However, there were many cases where the data collected by travellers failed to address key areas of ethnological inquiry. Responding to this, figures such as James Cowles Prichard, Thomas Hodgkin and Cull produced questionnaires that provided informants with instructions on how to acquire more detailed ethnographic observations. By the middle of the 1850s other ethnologists followed suit, and practitioners such as Robert Gordon Latham and Robert Knox aimed to further transform the observational practices of British researchers and informants by developing other techniques. Both of these figures produced didactic literature for ethnologists and travellers to utilize in their investigations, and they created local training sites in Britain for novice practitioners to develop specialized observational skills.

Because there was no consensus on how to use ethnographic evidence for ethnological investigations, during the 1850s there emerged two competing modes for studying human diversity.[2] The first was the more dominant monogenetic paradigm, based on the taxonomical system of Johann Friedrich Blumenbach as outlined in his treatise *De Generis Humani Varietate Nativa*. This book formed the basis of important ethnological texts such as the various editions of Prichard's *Researches*, as well as William Lawrence's lectures on the *Natural History of Man*.[3] The second explanatory model to gain considerable attention during the 1850s was polygenism, which argued in favour of the multiple origins of races. It had continental roots in French and German comparative anatomy, such as the taxonomical system of Georges Cuvier as found in his *Leçons d'Anatomie Comparée* (1800–5), and the transcendental morphology outlined in the works of Étienne Geoffroy de Saint-Hilaire.[4]

This chapter seeks to enlarge our understanding of early ethnological methods by looking at the observational practices of Latham and Knox. During the 1840s and 1850s both of these figures produced innovative techniques for transforming and refining the observational practices of British researchers interested in human diversity by establishing local sites for observational training. Latham and Knox are good examples for this inquiry because their contemporaries – as well as historians – have routinely compared their works and positioned them as representing opposing ethnological models.[5] Latham was framed as a monogenist and philologist, committed to advancing Prichardian ethnological methods. He also had close ties with the leading members at the ESL, including Hodgkin, Richard King and Henry Christy (1810–65). Knox, on the other hand, was an outcast of the ESL and a staunch polygenist, transcendentalist and racial determinist. He influenced a younger generation of practitioners – such as James Hunt and Charles Carter Blake (*c.* 1840–87) – by getting them to question some of the fundamental principles of the discipline, including the use of comparative language studies and racial theories based on religious dogma.

In this chapter, I will show that figures such as Latham and Knox were not passive observers and anthologists of ethnographic material, but important developers of the science. They were committed to producing new techniques for collecting ethnographic data, analysing materials and writing ethnological accounts. I will examine the vocational strategies of both figures in turn, and I will offer detailed analyses of Latham's first book *Natural History of the Varieties of Man* (1850) and Knox's famous work *Races of Man* (1850). Most examinations of Knox's text have emphasized his extreme racial and anti-Semitic views.

While these studies are elucidating, I want to move beyond his race discourse and focus on his observational practice. Knox saw training in anatomy and physiology as essential for scientific research, and he based his practical model both on his experience working as a military surgeon during the 1810s in Africa and on his educational background.[6]

In section two, I shift focus again and look at observational training grounds. Both Latham and Knox developed innovative techniques for directly engaging with their object of study by establishing local observational training sites in Europe. Knox visited Gypsy communities in northern England and Jewish communities in London and Amsterdam. By doing so, he could observe first-hand these populations *in situ*. Latham created an ethnographic training site at the Crystal Palace. Moreover, in 1854 he co-produced – with the naturalist Edward Forbes (1812–54) – a handbook entitled *The Natural History Department of the Crystal Palace Described*, which taught visitors how to observe the different races displayed within the department in a specialized way. For my purposes, I will examine the Zulu and Bushman installations.[7] In the third section I will look at a sample of periodical reviews of both Latham and Knox's books and discuss what their contemporaries thought of their works. I will show the growing divide emerging during the 1850s between their two competing methodologies.

## Competing Modes of Practice

> It might be called perhaps a minor curiosity in literature, that [two] ... metropolitan [MDs] ... should at the same moment have sent forth their researches on a subject so little known, or so imperfectly understood, as the varieties of the races of men, and that their several works should present the strongest possible contrasts to each other, contrasts so strong, that the terms light and darkness can alone define them.[8]
>
> *Bentley's Miscellany*, 1850

### *Robert Gordon Latham and the* Varieties of Man

Although Latham has received some recent scholarly attention, he remains a fairly obscure figure within the historiography. However, during the middle of the nineteenth century he was one of the leading figures within British ethnological circles. He began his education at Eton College in 1819 and entered King's College Cambridge in 1829, where he studied English language and literature, graduating with a BA in 1833.[9] After completing his degree, Latham spent a year travelling and living in Hamburg, Copenhagen and rural Norway. His time on the continent had a significant impact on his future ethnological activities, and it was during this period that he discovered his passion for philology and metagraphy.[10] Latham spent many hours learning Danish and Norwegian, and one nineteenth-century commentator stated that during his time abroad, 'his reading was largely, if not exclusively, philological'.[11]

The time of Latham's immersion into Scandinavian literature and culture was an important developmental period, and it afforded him with a form of first-hand ethnographic experience. It also shaped his later interests in the study of human diversity. There were certainly striking differences between early nineteenth-century Denmark and rural Norway, and the late Georgian society from which Latham came. His knowledge of Danish and Norwegian culture also featured in his later ethnological work on Europeans.[12] Upon his return to Britain in 1834, Latham took up an MA with a focus on language and literature at Cambridge, which he completed in 1836. While he worked on his MA he began to study medicine, and he qualified as a licentiate of the Royal College of Physicians in 1842 and received his MD from Cambridge in 1844.[13] Latham's educational background is clearly reflected in his ethnological writings, and in his first major book *Varieties of Man*, he examined both the physical structure of different ethnicities and their languages.[14] He based his ethnological framework on the Prichardian model, and in the preface he praised Prichard's contributions to the study of human variation:

> If the simple excellence of a book were a sufficient reason for making it the only one belonging to the sciences which it professed to illustrate, few writers would be desirous of attempting a systematic work upon the Natural History of their species, after the admirable [Researches into the] Physical History of Mankind, by the late ... Dr. Prichard ...[15]

This was an astute move and almost certainly linked to Latham's vocational strategy. The British ethnological community was looking for a new disciplinary figurehead to carry the science forward after Prichard's untimely death in 1848. Latham's commitment to further transforming and refining Prichardian ethnology endeared him to the senior members of the ESL – such as Hodgkin and Christy – and enabled him to rise to prominence as a leading scholar within the research field.

Latham argued that the *Researches* was 'a work of rising rather than stationary science; and the very stimulus which it supplied, created and diffused a spirit of investigation'.[16] As with Prichard and other early nineteenth-century ethnologists before him, Latham believed that the ability to identify different human varieties based on their physical and cultural attributes was central for ethnological inquiries. His *Varieties of Man* was a taxonomy of human diversity, and it taught practitioners how to observe human variation in a specialized way. Using Latham's method, readers could identify different human varieties based on their physical form, geographical location and language. In his introduction, Latham argued that 'Previous to entering upon the details connected with the varieties, and affinities of the human species, it is advisable to explain the meaning and full import of certain terms that are likely to be of frequent occurrence'.[17] He began by outlining the three main varieties of humans, which were 'the Mongolian', 'the Negro' and 'the European'.[18] In each case Latham provided his readers with a comprehensive breakdown of each type, with a particular focus on their facial structures. He wrote,

> The Mongolian's ... face would be broad and flat, with the cheek-bones prominent. The breadth of the head from side to side would be nearly equal to its length from the forehead to the occiput; the nose would be flat, and, almost certainly, neither arched nor aquiline; the eyes would be drawn upwards at their outer angle, the skin would be of a yellowish-brown, the hair straight, the beard scanty, and the stature undersized.[19]

He based these descriptions on Blumenbach's explanatory analysis of human variation from *De Generis Humani Varietate Nativa*, and this was a continuation of earlier ethnological approaches. Prichard and Lawrence had also built their classificatory systems on Blumenbach's work.[20]

A medical education was useful (but not necessary) when learning Latham's taxonomical method, and training in anatomy and physiology furnished readers with specialized observational skills for examining human diversity. For Latham, the most important physical feature – which separated human varieties – was the conformation of the skull. As such, skin, eye and hair colour were a secondary concern. He stated that 'Although the colour of the skin and eyes and the texture of the hair may be determined by external influences, the real reasons for the differences of *outline* lie in the differences of the skull and the bony parts of the face: and as, in addition to this, the skull is the receptacle of the brain, and the brain is the organ wherein the human species most differs from others, anatomists have long been in the habit of determining the different varieties of the human race, by the difference in the conformation of their skulls'.[21]

Because it was the conformation of the skull that primarily differentiated human varieties, Latham included descriptions of the key bones researchers were to examine when studying different ethnicities. They included the 'Frontal Bone', the 'Occipital Bone' and the 'Parietal Bones'.[22] The descriptive terminology was highly technical, and it would have been difficult for non-medical specialists to follow without some knowledge of anatomy. For example, Latham discussed 'Parietal Bones' as follows:

> *The Parietal bones, forming the sides of the skull.*– The distance between the two parietal bones is the parietal diameter. It constitutes the breadth of the skull, in contradistinction to its *length* or *depth* ... Nations where the development is in the occipito-frontal diameter are called *dolikhokephalic*. Nations where it is in the *parietal* diameter are called *brakhykephalic*.[23]

It was important that the reader learned to develop their discriminating comparative anatomical skills, and to assist them with locating these features on skulls, Latham included a diagram of two different types of skull conformations representing both the 'dolikhokephalic' and the 'brakhykephalic' forms (see Figure 3.1).

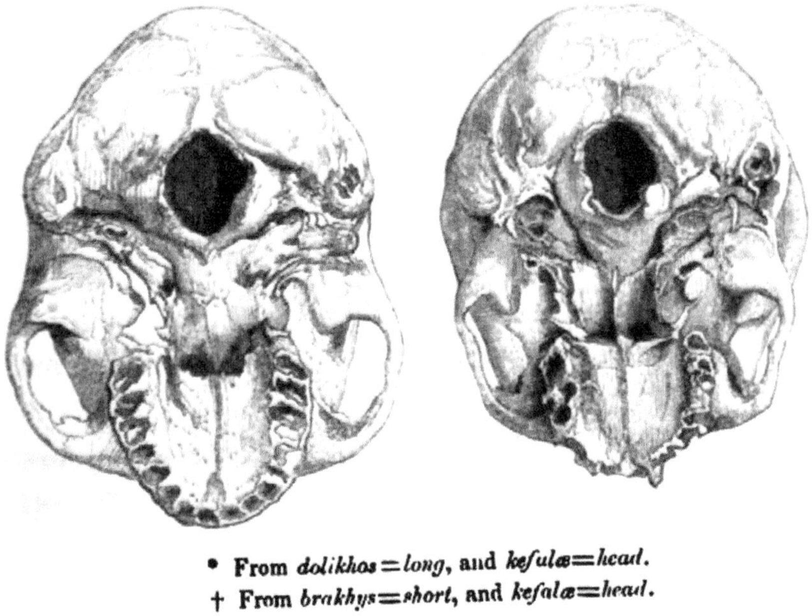


**Figure 3.1: A diagram of 'parietal bones' from Latham's *Varieties of Man*. From R. G. Latham, *The Natural History of the Varieties of Man* (London: John Van Voorst, 1850), p. 5. Reproduced from the author's private collection.**

In addition to basing his taxonomical categories around designations that emphasized differences in skull conformation, Latham argued that it was useful to maintain the traditional classification titles based on skin, eye and hair colour. He wrote, 'the terms denoting the chief divisions have not been given upon anatomical grounds. Hence we do not talk of the *zygomatic* or the *occipito-frontal* tribes, but the *Negro*, or the *Mongolian*, &c.'[24] This was helpful for readers who were familiar with Blumenbach, Prichard and Lawrence's works because it created some continuity between the various texts. Moreover, it made the taxonomy slightly more accessible to researchers with limited or no anatomical training – such as colonial officers working abroad – because it was faster to identify different varieties by external features such as skin colour. Latham was uncertain about accepting Blumenbach's argument for five main varieties of humans, and he stated that the 'Malay' and 'American' were part of the other three primary ethnic groups. He wrote, 'How many of such terms are necessary is a disputed point; the number of the primary divisions being underdetermined. My own opinion is in favour of it being limited to *three*, – the Mongolian, the African, and the European'.[25]

Once Latham established that he based his classification around three main varieties, he proceeded by further refining his categories and explaining the terms 'type' and 'sub-type', which he utilized frequently within his taxonomical vocabulary: 'Now, so far as this difference is concerned, the terms *typical* and *sub-typical*, in their usual sense, are sufficient; the Mongolian being the *type* of the variety which he represents, whilst the Malay and American each illustrate a *sub-typical* modification'.[26] Within sub-types there were two forms, 'simple' and 'deviation with transition', and distinguishing between these variations was complex. Thus Latham outlined a method for identifying the difference between the two kinds:

> 1st. *Simple sub-typical deviation.* – This occurs in the Eskimo. His face is broader than that of the Mongolian, but, as the increased breadth merely makes him somewhat unlike the native of Central Asia, without approximating him to the African or European, the deviation is *simple*.[27]
>
> 2nd. *Deviation with Transition.* – The Finlander has a Mongoliform skull, but (very often) blue eyes and light hair; so that he agrees with the European where he differs with the Kalmuk. This deviation is something more. It is deviation accompanied with the phenomenon of a transition in form.[28]

Sub-types, which represent 'deviation with transition', had two kinds: first, 'those which descent plays a part (a)'; and second, 'those in which causes other than descent play a part (b)', for instance, the effect of climate and environment on the physical form of a human variety.[29] Latham elaborated on both kinds:

> *a.* The light haired Finlander is probably one of three things –
> 1. The descendant of Mongolians passing into Europeans
> 2. The descendant of Europeans passing into Mongolians
> 3. The descendant of the common stock from which the Europeans on one side, and Mongolians on the other, originated. In all these cases his differential characters are accounted for by the doctrine of *descent*[30]
>
> *b.* Contrast, however, the case of an Australian Black. He has Mongol characters and he has Negro characters; so that, looking at his *form* only, he presents the phenomenon of transition; yet he is in none of the predicaments of the Finlander, since few ethnologists believe that, *in any way of descent*, he has any but the most indirect relationship to the African[31]

Latham's method for identifying and organizing different human varieties into groupings based on similarities and differences had clear links to both Prichard's and Lawrence's taxonomies from the early part of the nineteenth century, and Blumenbach's system from the eighteenth century. However, Latham's classification was much more refined than the earlier versions; he not only categorized human difference by 'type', but took the method one step further by identifying more subtle details and placing different ethnicities into 'sub-types'. Latham also incorporated his training in linguistics, metagraphy and philology into his framework and – as with Prichard – he argued that through language the ethnologist could determine common links between different human varieties. He wrote, 'there are three methods by which the relation between the different words that constitute sentences is indicated: – 1. The method of which the Chinese is a sample; 2. The method of which the Greek and Latin are samples; 3. The method of which the English is a sample'.[32] The researcher was to pay close attention to inflections in words and compare the structure and pronouncement of sentences. Latham believed that ethnologists could advance monogenetic arguments by identifying the etymological origins of different dialects to specific regions of the world. He wrote for instance, 'Where a Roman said *te-tig-I*, we say *I have touched*, or *I touched*; using *I*, a separate word, instead of the incorporated syllable *–i*. Where a Roman said *part-i*, we say *to father*; where a Roman said *tang-am*, we say *I will* (or *shall*) *touch*. In other words, we make auxiliary verbs and prepositions do the work of inflections, expressive of case and tense'.[33]

Once Latham defined the chief classificatory terms within his model and explained his method for grouping humans into primary types based on their physical structure, he proceeded by further refining his taxonomical system and dividing humans into geographical regions. Under this model, all humans belong to both 'divisions' and 'sub-divisions'. For instance, within the 'Mongolian' variety there were seven divisions, 'Altaic', 'Dioscurian', 'Oceanic', 'Hyperborean', 'Peninsular', 'American' and 'Indian'. Within each of these divisions and sub-divisions, Latham further grouped the inhabitants into 'tribes' and 'nations', with brief descriptions of their physical form, geographic location and language. For instance, the 'Tasmanians' inhabited Van Diemen's Land,[34] and he described their physical form as 'Negritos, with curly, frizzy, or woolly hair; *i.e.*, with the character of the Papua, but not within the Papua geographical area'.[35] Latham further stated that

> Writers who are not, otherwise, over-prone to exaggerate differences, have separated the Tasmanians from the Australians; and this arrangement is followed in the present work. The physical difference is chiefly that of the hair. The language, as far as the imperfect vocabularies have allowed me to examine it, has fewer affinities with the southern dialects of Australians than even the known amount of dissimilarities between fundamentally allied languages prepares us for.[36]

He also included an image of a 'Tasmanian' (see Figure 3.2), which was useful for researchers who wanted to develop their comparative observational skills, as it afforded them with an opportunity to contrast Latham's written descriptions with an illustration.

**Figure 3.2: An image of a Tasmanian from Latham's *Varieties of Man*. From R. G. Latham, *The Natural History of the Varieties of Man* (London: John Van Voorst, 1850), p. 245. Reproduced from the author's private collection.**

Because the population of Tasmanian aborigines had been greatly diminished from diseases they contracted through cross-cultural encounters with Europeans, there was a fear that their culture was dying out. Early ethnologists wanted to collect as much information on these people as possible. Paintings and sketches were equally important, and by the 1850s visual representations of Tasmanians were significant for ethnologists because many believed that these images were the only remaining depictions of these aborigines.[37]

Latham's *Varieties of Man* aimed to teach researchers how to identify and organize different humans into groupings based on their physical form, geographical location and language. He continued to transform and refine the observational practices of researchers interested in human variation. However,

Latham's practical framework for studying human diversity was not the only approach available for researchers during the middle of the century, and a competing mode gained momentum during the 1840s and 1850s. Knox was one of the earliest British figures to forward this divergent methodology.

## *Robert Knox and the* Races of Man

In the first half of the nineteenth century, one of the most controversial and outspoken British practitioners of the sciences of Man was the prolific lecturer and writer Robert Knox. He began his career during the 1820s as a leading anatomist in Edinburgh. However, after his involvement with William Burke (1792–1829) and William Hare in the West Port murders,[38] Knox's reputation as a respected man of science dwindled.[39] Nevertheless, this is not another examination of the impact of the Burke and Hare scandal on Knox's career. Nor is it a study of the influence his writings had on racial determinism, although it will be discussed briefly. Rather, it is my intention to show another side of Knox, which focuses on his contributions to ethnological and anthropological observational practices. Throughout his career he was widely recognized as one of the greatest anatomists of his day, and his students and followers – such as Hunt, Blake and Henry Lonsdale (1816–76) – praised his abilities as a scientific observer.[40]

There is a scarcity of archival material remaining because Knox destroyed most of his letters and manuscripts before his death in 1862. Even the materials that Lonsdale used in writing Knox's biography are lost.[41] What remains is Lonsdale's account from 1870 and Knox's published works, which for the most part he wrote after 1842, when he left Scotland deeply embittered. Understanding Knox is further complicated because of the scattered and disjointed nature of his writings. During the 1840s and 1850s he barely lived above a level of pauperism, and he wrote most of his books in haste because of his desperate need for money. As a result, it is not a straightforward task to make sense of Knox's manifesto, and in order to grasp the complexity of his observational practices, the historian has to have a familiarity with several of his books. A comprehensive understanding of his analytical framework is only achieved by bringing together components from several different sections of his published works. As was the case with many of the figures we have already examined, Knox was not a 'passive observer' of ethnological and ethnographic knowledge, but rather, he was actively involved in collecting data and shaping its meaning and production.

Knox's intellectual abilities were evident from a young age. In 1805 he entered Edinburgh High School, where he was a dux and gold medallist.[42] After graduation he enrolled in the medical school at the University of Edinburgh, where he quickly rose in popularity with his medical peers, becoming the president of the Royal Medical Society twice before his graduation as an MD in 1814.[43] Interestingly, during this period of study at the university, Knox failed his first examination on human anatomy – a subject in which he became a leading British expert by the mid-1820s, and which formed the foundation of his ethnological framework. It was under the so-called incompetence of Alexander Monro Tertius (1773–1859) that Knox failed to initially acquire a thorough understanding of the physical structure of humans. Thus between 1813 and 1814 he studied anatomy at John Barclay's extramural school in Edinburgh.[44] Under the tutelage of the renowned anatomist John Barclay (1758–1826), Knox mastered the subject, and in 1814 he passed his medical exam at the university.

Knox's earliest encounter with extra-Europeans occurred between 1817 and 1820, when he was stationed as a military surgeon during the Cape Frontiers War in South Africa with the 72nd Highlanders. Inside the hospital, Knox gained a sound knowledge of human anatomy and physiology operating on European and African casualties of war.[45] Outside the hospital, he gained invaluable natural history experience. His student and biographer Lonsdale recounted that Knox's experience in Africa had a significant impact on his medical and scientific practices because it afforded him an opportunity to observe *in situ* natural history specimens.[46] He wrote,

> As a man of thought and observation, Knox looked beyond the confines of his hospital engagements, and aimed for the light which fresh countries and travel afford. He examined the charts of the Cape territory, and was able to correct and extend them ... He studied the physical geography and meteorology of each district, and not less the Fauna, if not the entire natural history under his survey.[47]

The Cape provided Knox with a rich location to observe humanity because several different ethnic communities, including Europeans and Africans, came into contact with each other. Lonsdale wrote that 'Early in his African experiences the ethnological faculty seemed strongly manifested in Knox',[48] and he meticulously recorded the physical and cultural traits of the indigenous people he encountered. This material was incorporated into both his lectures on human variation and his published material on racial groups. Lonsdale wrote, 'a great deal of what he saw in Africa came to be worked up in magazine articles, whilst the anthropological *memorabilia* formed interesting details for his anatomical classes, and were afterwards woven in graphic colours with his "Fragment" on the Races of Men'.[49]

Knox examined at length the anatomy and physiology of the various peoples that populated the earth in his *Races of Man*, and he also discussed different cultural practices that were specific to each group. The most significant aspect of his book, which separated his study from the work of his contemporaries, was that he based several of his racial descriptions on his own first-hand observations taken *in situ*. For example, Knox substantiated his account of the 'Hottentots' and 'Bushmen' with a mixture of data taken from his own primary experiences living among these indigenous groups and from the testimonies of travellers.[50]

This was considerably different from the work of other ethnologists from the period – such as Prichard – who were solely reliant on the observations of informants to supply them with ethnographic data for their investigations.

Knox's theoretical perspective differed extensively from the Prichardian model, and for him race was everything. It not only shaped a person's physical form but also their social behaviour. He recognized that such a stance placed him in staunch opposition to many of the leading ethnologists at the ESL, and he wrote in his preface that 'The "Fragment" I here present to the world has cost me much thought and anxiety; the views it contains being so wholly at variance with long received doctrines, stereotyped prejudices, national delusions, and a physiology and philosophy, if it can be so called, as old at least as the Hebrew'.[51] He continued by stating that

> The human character, individual and national, is traceable solely to the nature of that race to which the individual or nation belongs, is a statement which I know must meet with the severest opposition. It runs counter to nearly all the chronicles of events, called histories; it shocks the theories of statesmen, theologians, [and] philanthropists of all shades ...[52]

Knox was ruthless in his attack on traditional ethnological practice, and he referred to figures such as Prichard and Lawrence as 'hack compilers' because they based their studies on travellers' accounts. He also argued that these ethnologists were committed to 'false doctrines' – such as monogenism – and he believed that his opponents would omit his divergent theories and accounts from their work because it would discredit their research.[53]

In his introduction, Knox presented a vastly different approach to the study of human diversity by forwarding a polygenetic view heavily influenced by comparative anatomy and transcendentalism. He wrote that 'Of man's origin we know nothing correctly; we know not when he first appeared in space; [and thus] his place in time, then, is unknown'.[54] He continued by attacking the chronologies of naturalists who attempted to outline the course of human history, and he stated, 'how worthless are these chronologies! How replete with error human history has been proved to be'.[55] For Knox, origin theories were a secondary concern for any scientific study on humans. Instead, the primary aim for researchers was to observe and describe the current form of different races. French naturalism was central to Knox's methodological approach.

In 1821 Knox had moved to Paris, where he studied comparative anatomy with Cuvier and transcendentalism with Saint-Hilaire. Knox's time in France was an important developmental period that shaped his future theoretical and methodological frameworks. For Knox, comparative anatomy provided researchers with a practical method for organizing human varieties into groupings based on similarities and differences, while transcendentalism furnished them with a basis for all theories on races. In his book *Great Artists and Great Anatomists* (1852), Knox praised the contributions of both Cuvier and Saint-Hilaire to medicine and science.[56] He wrote that Cuvier's comparative anatomy formed 'the basis of all zoological knowledge ... in other words, to the science of organic beings'.[57] Knox reiterated the centrality of anatomical training in many of his books, and he wrote in his *Manual of Human Anatomy* (1853) that

> The useful arts of Medicine and Surgery are only entitled to the name and dignity of professions when based on descriptive Anatomy. It is to the labours of the anatomy of adult forms that Medicine, Zoology, and Geology owe their present position, whatever that may be. For the Medicine which appeals not to Anatomy is empiricism; and the Zoology not based thereon is not a science.[58]

Transcendentalism was equally essential for Knox, and he believed there was an underlying structural relationship between diverse groups of organisms in nature. His transcendentalism was a primary component of his theoretical framework for studying races, and he wrote that 'transcendental anatomy, which alone of all systems, affords us a glimpse and a hope of a true "theory of nature"'.[59] Thus Cuvier and Saint-Hilaire both had an impact on Knox's writings regarding human variation. It was evident that the physical structure of different human varieties was the cornerstone of his ethnology, and he wrote in *Races of Man* that

> The basis of the view I take of man is his Physical structure; if I may so say his Zoological history. To know this must be the first step in all inquiries into man's history: all abstractions, neglecting or despising this great element, the physical character and constitution of man, his mental and corporeal attributes must, of necessity be at the least Utopian if not erroneous.[60]

For Knox, the key to studying different races lay in the observational abilities of the practitioner, and he did not want to overemphasize the importance of natural history terminology. It was more important that the researcher learned to observe and identify differences in physical form based on anatomical and physiological training, than it was to organize different populations into varieties based on natural history taxonomical vocabularies. He repeated this point in many of his works, and for instance he wrote in *Great Artists and Great Anatomists* that 'there are schools and colleges to which men have been appointed to mislead and misdirect the youthful mind; to teach him to substitute for his own tact and powers of observation ... a [terminological] definition for actual observation through the senses'.[61]

In *Races of Man*, Knox argued that 'Men are of various Races; call them Species, if you will; call them permanent Varieties; it matters not. The fact, the simple fact, remains just as it was: men are of different races'.[62] He continued by attacking Blumenbach's and Prichard's taxonomy of human varieties, and he stated that 'I early examined the work of Blumenbach, of which the laborious writings of Dr. Prichard were an extension – [as] an imperfect work, leading to no results; teaching a physiology as old as Herodotus and Hippocratus'.[63] Knox continued his aggressive critique by remarking that 'the philosophic formula of Blumenbach led to no results: explained nothing: investigated no causes. It was the external-character naturalist trying "his method" on man'.[64] Responding to this, Knox argued that different races were separated not only by their external anatomical features, but also by their internal physiology. He believed that taxonomies based on Blumenbach's system were inherently flawed because they ignored this aspect of the human physiognomy. Knox wrote that Blumenbach's method 'left every great physiological question unanswered'.[65]

Knox wanted to promote a racial theory which argued that all humans were biologically determined. Therefore all non-physical attributes were dependent upon physical conditions. He argued, for example, that if Africans were educated in Britain and enculturated into European society, their intellectual abilities and social position would never exceed the 'natural capabilities' of their race. Even if their immersion into a culture influenced their behaviour, he still believed they would ultimately remain biologically distinct. He wrote, 'If any one insists that a Negro or Tasmanian accidentally born in England becomes thereby an Englishman, I yield the point; but should he further insist, that he, the said Negro or Tasmanian, may become also a Saxon or Scandinavian, I must contend against so ludicrous an error'.[66]

This was a markedly different analytical framework to the monogenism of figures such as Prichard and Lawrence because Knox viewed all humans as distinct from one another: 'With me, race, or hereditary descent, is everything; it stamps the man. Setting aside all theories, I have endeavoured to view mankind as they now exist, divided as they are, and seem always to have been, into distinct races'.[67] Thus a divergent ethnological model gained momentum within British scientific circles at the start of the 1850s, and Knox was a chief figure in promoting this interpretive system. Moreover, during this same period, Latham and Knox were both developing other innovative techniques for transforming the observational practices of British researchers. Latham created a training site at the Crystal Palace exhibition, while Knox conducted a form of fieldwork by visiting ethnic communities in the Netherlands and Britain to gain first-hand knowledge of different races.

## Local Sites of Training

> The science of ethnology is, to a great extent, a new one, and this has been our excuse for enlarging upon the meaning, and entering into the origin of the word. Even ethnological museums are rare.[68]
>
> Robert Gordon Latham, 1854

### *Latham and Ethnology at the Crystal Palace Exhibition*

In 1852 the Crystal Palace Company appointed Latham as director of the ethnological department, and in 1854 the exhibition relocated to Sydenham. The new building was designed by Joseph Paxton (1803–65), and it was twice the size of the original site and possessed larger collections. In the same year Latham published a small guidebook that described the content of the ethnological exhibit.[69] It was part of a larger text co-edited by Latham and the naturalist Edward Forbes, which described the entire natural history section at the Crystal Palace.[70] The guidebook had several uses. It provided the general public with a detailed account of the ethnological department's collections. This was also useful for interested parties unable to visit the Crystal Palace. For a more specialized visitor wanting to refine their observational practices, the text enhanced their visual experience because it explained to readers/visitors how to observe and interpret the different human varieties on display in a specialized way. The ethnological department was therefore a type of ethnographic training ground. The department was also significant for ethnology because it afforded the discipline an opportunity to show its value for the empire. Latham was charged with the duty of displaying the various races that inhabited British territories. By doing this, he was providing useful information about these peoples. The Crystal Palace exhibition enabled Latham to demonstrate the important role ethnologists played in disseminating ethnographic information to British audiences. It also strengthened ethnology's partnership with colonial officials by emphasizing the reciprocal benefits of educating Britons about their territorial wealth and the people who inhabited it.

Bringing together specimens from ethnology, botany and zoology, Latham attempted to recreate scenes depicting different human types in their natural environments. He wrote, 'The trees, plants, animals, and human occupants of the different portions of the earth's surface are grouped together – so that the allied sciences of botany, zoology, and ethnology illustrate each other'.[71] The exhibit afforded researchers with an opportunity to practise their observational techniques on realistic-looking ethnographic displays because it created a local site in the metropole where ethnologists could visualize extra-European peoples in a simulated natural habitat, without having to travel to distant locations scattered throughout the empire. It was also an important ethnographic resource for informants – or anyone else interested in potentially contributing to both ethnographic and ethnological research – because it provided them with an opportunity to refine their observational skills before embarking on their voyages. Members of the scientific community acknowledged the importance of the ethnology department at the Crystal Palace. For instance, after visiting the exhibit, the ethnologist and physician John Conolly (1795–1866) wrote that the department was arranged 'with a view to ... greatly facilitate the progress' of science.[72] Even after its destruction by fire in 1867, ethnologists continued to comment on the significance of the site as an educational training ground. James Hunt remarked that the exhibit contained 'the only material generally accessible to the public in London, by which the popular mind could render itself familiar with the aspects of many of the races of man'.[73]

Latham organized the different ethnographic displays geographically (see Figure 3.3) so that visitors could find their way around the department as if they were circumnavigating the globe. He wrote, 'The arrangement is so far geographical that, to a certain extent, the visitor is enabled to place himself in respect to the objects before him in the same relation as he would be to a map of the world'.[74] In each of the regions of the world exhibited, Latham provided his readers/visitors with an abundance of information about the humans on display. It was similar to how he examined different tribes and nations in his *Varieties of Man*. He discussed their physical form, geographic location, language and social behaviour. Latham also identified for readers/visitors which main physical and cultural features to focus on, and instead of simply seeing extra-European peoples in foreign attire and surrounded by strange plants and animals, he transformed readers/visitors into skilled observers.

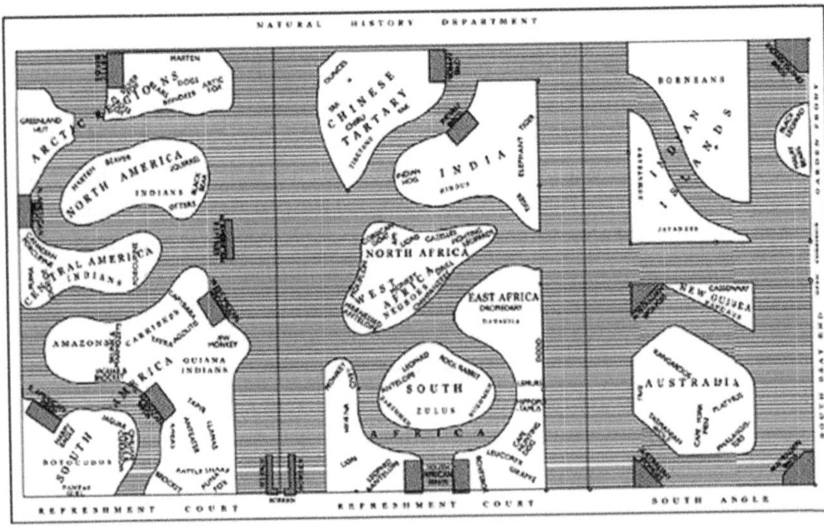


**Figure 3.3: A map of the ethnological exhibit at the Crystal Palace. From S. Phillips, *Guide to the Crystal Palace and its Park and Gardens* (London: Bradbury and Evans, 1854), pp. 116–17. Reproduced from the author's private collection.**

The accuracy and authenticity of the ethnological specimens were integral for Latham. However, the ethnological department did not have living representatives of each of the various cultures on display. Instead, each installation had realistic-looking mannequins painted with the same skin pigmentation colour as the extra-European populations they represented (see Figure 3.4). In addition, Latham dressed each model in their supposed indigenous clothing and fitted them with face-casts of real indigenous people acquired *in situ* by informants travelling and living abroad.[75] In the case of British Guiana, Latham wrote, 'These are from casts taken from life during Sir. [Robert Hermann] Schomburgk's expedition. All belong to the great Carib stock'.[76] In the case of the 'Bushmen', Latham wrote, 'The Bushmen, too, are taken from life, the two children being in England at the present time'.[77]

**Figure 3.4: The 'Bushmen' display from the Crystal Palace. Reproduced by permission of the Pitt Rivers Museum, University of Oxford, catalogue no. PRM 1998.211.9.**

Latham also provided detailed descriptions of each race; for example, he stated that the 'Bushmen' belonged to the 'Hottentot' family and that they were located 'on the *western* rather than *eastern* side of South Africa, and extend from the parts about Valvisch Bay to the Cape'.[78] In terms of their physical form, he wrote, '[The Bushmen] ... are described as having constitutions "so much enfeebled by the dissolute life they lead, and the constant smoking of *dacha*, that nearly all, including the young people, look old and wrinkled; nevertheless, they are remarkable for vanity, and decorate their ears, legs, and arms with beads, and iron, copper, or brass rings ...".[79] Latham also noted that 'The women likewise stain their faces red, or paint them, either wholly or in part. Their clothing consists of a few sheepskins, which hang about their bodies, and thus form the mantle or covering, commonly called a *Kaross.* This is their only clothing by day or night'.[80] If readers/visitors wanted further information on 'Bushmen' – or any of the other ethnic groups on display in the department – they could read Latham's lengthier analyses from the *Varieties of Man.*

In cases where Latham arranged some mannequins in order to depict a cultural practice, he explicated to readers/visitors the cultural significance of the ritual they observed.[81] In the case of the 'Zulu' display (see Figure 3.5), he wrote the following:

> The story told is the search for some lost article. When this is the case, a Fetish-man, medium-man, mystery-man, or conjurer (we may choose our name), is called in, and set upon the suspected parties, who sit round in a circle. The conjuror then works himself, like Pythoness of the old oracles, into a state of rabid excitement, and keeps it up until he fixes upon the culprit.[82]

**Figure 3.5: The 'Zulu' display from the Crystal Palace. Reproduced by permission of the Pitt Rivers Museum, University of Oxford, catalogue no. PRM 1998.210.3.**

With this description, readers/visitors learned the significance of the scene they observed before them. Latham argued that travellers had witnessed similar rituals in other areas of Africa, and 'Fetish-men' used similar techniques to detect crime in other societies. This was important information for ethnologists and informants for several reasons. It demonstrated that comparable cultural practices existed among different ethnic groups, and this fed into monogenetic arguments because if two distantly located communities shared analogous social behaviours, the evidence could be used to support the possibility of a single ancestral origin. Conversely, if an informant stumbled upon this cultural practice *in situ*, they now possessed some knowledge of what they were witnessing. This in turn would produce superior ethnographic reports because informants knew what aspects of the ritual to record for ethnologists back in Britain.

Latham transformed the ethnographic exhibit at the Crystal Palace into a training site for researchers and potential informants. Using his guidebook and other ethnological literature such as his *Varieties of Man*, visitors could learn how to observe the various ethnic groups on display at the department in a specialized way. It is significant that the surviving photographs from the Chrystal Palace exhibition come from Edward Burnett Tylor's personal collection.[83] He would have acquired these images by visiting the department. It is suggestive that he held onto these photographs because feasibly he viewed the ethnological display at Sydenham as an important ethnographic site for early practitioners. Latham was not the only ethnologist in Britain developing sites locally for ethnographic observational training. Robert Knox also encouraged practitioners to utilize locations in Britain and Europe as a way to improve their observational techniques and researches.

## *Knox's Observations of Gypsies and Jews*

For Knox, the key to studying different races lay in the observational abilities of the practitioner. He argued that it was essential to have primary experience of observing humans *in situ*. He rejected scientific inquiries that substantiated their claims solely based on the use of second-hand information, believing most secondary accounts were biased and flawed. Lonsdale noted in his biography of Knox that

> In all that related to fact and observation in reference to or in illustration of science, Knox was scrupulous to a degree that nothing but the truth should be emitted. In describing the results of his own investigations, he was ever anxious to keep prominently in view the tendency that preconceived notions … had in swaying inferences, or in distributing the fair expressions of a fact.[84]

According to Knox, ethnological accounts were to be based primarily on direct observation that was free from the social and political agendas of researchers. The ethnologist was to describe the physical and non-physical attributes of a race without applying any preconceived theories that would influence their analysis of the data. Secondary reports by informants were to be used with discretion.

Any account of a human variety collected by an informant without training in anatomy and physiology was discreditable. Knox argued that 'When some superficial observer has made a few remarks on the skeleton of a race, he fancies he knows its anatomy!'[85] Consequently, for Knox, it was integral for ethnographic research to be conducted by trained practitioners. His insistence on having anatomical and physiological training almost certainly linked to his vocational strategy. Knox struggled for employment after he was publicly defamed for his involvement in the West Port murders. The only way he was able to make a living afterwards was by delivering lectures and publishing books on human variation topics. Therefore, by positioning medically trained researchers such as himself at the forefront of ethnological research, he was attempting to increase his ticket and book sales and gain a leading role as a chief expert within the budding discipline.

Knox's insistence on ethnology being based fundamentally on first-hand experience is significant for others reasons. It demonstrates a common link between some mid-Victorian research methodologies and early twentieth-century fieldwork-based anthropology. Knox argued that training in anatomy and physiology was essential for 'skilled observation', and this was comparable to Malinowski's emphasis during the 1910s on the importance of having university-trained ethnographers conduct anthropological research. However, even with these similarities in mind, there are also marked differences between ethnography in the 1850s and its counterpart in the early twentieth century. Most notably, nineteenth-century ethnologists were not using a rigorous form of participant observation in their studies, as was the case with later figures such as Malinowski. Moreover, Victorian practitioners did not prioritize locating isolated communities on remote islands as essential for their research; exoticism was not a necessary attribute for all Victorian ethnology. Ethnologists in the middle of the nineteenth century were equally committed to producing comprehensive ethnographies of Europe.[86]

Knox's desire to base his work on primary experience led him to develop other innovative observational techniques. For example, he utilized local sites in Britain and Europe for conducting ethnographic field-based research. Knox was constantly collecting his own ethnographic data, and his biographer Lonsdale noted that

> Knox seemed to the manner born to investigate distinctive anatomical characters: even when walking along the streets, thronged with men and women, he was always on the *qui vive* for Race features. He could see at a glace what ordinary men could hardly distinguish at their leisure; if his eye was penetrative beyond most, his more gifted vision lay within an alert and discriminative mind.[87]

The observations that Knox collected while travelling across rural Britain, and roaming the streets in cities such as London and Amsterdam, furnished him with ethnographic data for his analyses of both the Gypsies and Jews. Unmistakably, his accounts of both groups are filled with racist and anti-Semitic comments. Nevertheless, they also promoted an innovative technique for conducting ethnographic research locally without having to travel to exotic locations such as Africa or the South Pacific. Most importantly, this meant that the researcher could record their own observations and not rely exclusively on second-hand reports. Knox opened his account of the Gypsies by stating that he based his racial descriptions on his first-hand experiences visiting their communities in the middle and north of England. He wrote, 'Let me state to you calmly the facts I have myself witnessed, the few observations I have made on this race'.[88]

Knox continued by recounting some of his experiences living among British Gypsy communities, and interestingly, he discussed both the physical characteristics of the people he encountered and their cultural activities and social organization: 'In the autumn of 1846, I resided for a considerable time in Derbyshire ... [and it was] infested with gipsy gangs'.[89] Knox described the physical structure of male Gypsies as 'well enough made, small and active', and he wrote that female gypsies physically 'look well for a short time, but they have not the elements of beauty, or at least very few of them'.[90] In terms of their clothing, Knox stated that the Gypsies were 'never well dressed – they and their children are in rags; the middle-aged men, on the contrary, are generally well dressed, well shod, comfortably arranged in all their apparel'.[91] In terms of their language, he wrote that it could be traced to 'the roots of many Hindostanee [*sic*] words, and they are obviously an Eastern race'.[92]

Another ethnic group that Knox visited and collected ethnographic data on was the Jewish communities in London and Amsterdam. As with his description of Gypsies, Knox discussed both the physical and cultural characteristics of Jews. He wrote, 'I had reached London, that compound of all the earth, and I had looked attentively at the Jewish physiognomy on the streets, as he perambulates our pavements, and with a hoarse, unmusical voice, proclaims to you his willingness to purchase the cast-off clothes of others; or assuming the air of a person of a different stamp, he saunters about Cornhill in quest of business'.[93] Using his training in anatomy and physiology, Knox argued that he observed three distinct Jewish varieties, and he stated, 'As I attentively surveyed the Jewish population on the streets of London, I fancied I could perceive three different casts of features: the first Jewish, *par excellence*, and never to be mistaken; a second, such as Rembrandt drew; and a third, possibly darker, of races intermingled'.[94] However, Knox was concerned that the Jewish populations he encountered in London were not prototypical because in many instances he believed British Jews had intermarried and procreated with other ethnicities. He decided to travel to Amsterdam and compare Dutch Jewry with the data he collected in London.[95]

Knox arrived in Amsterdam and visited the Jewish neighbourhood. He wrote that 'Having repaired to the quarter of the city occupied by this race in Amsterdam, I found the synagogue open and crowded; divine worship was going on, the people standing in crowds around the high altar; it was not proper to take off the hat'.[96] Knox sat in the synagogue and observed the activities of the worshippers. He also acknowledged some of the rudimentary social divisions that existed within Jewish society, and he noted that inside the synagogue the women sat in a separate section from the men.[97] Knox's followers such as Blake recognized the ingenuity of his observational technique, and he wrote that

> The principles which Knox laid down were applied by him to the Jewish race, and we must commend all Knox's generalizations respecting the Jews to the attention of the present generation of anthropologists ... they form the groundwork on which we hope that some future anthropologist will base a reliable memoir on the physical, moral, and mental history of the Jews.[98]

Knox's innovative technique of using local sites for observing human varieties *in situ* was commended by some of his contemporaries and incorporated into their own ethnological/anthropological practices. However, there were also many practitioners who criticized his methods and theories. Thus looking at periodical reviews of Latham's and Knox's books respectively is elucidating because it shows how their contemporaries interpreted their research programmes.

## Reviewing Latham and Knox

The diverging methods for studying human varieties outlined within both Latham's and Knox's paradigms met with markedly different responses within periodical reviews. For the most part, reviewers praised Latham's work and harshly criticized Knox. One of the earliest periodicals to discuss Latham's *Varieties of Man* was the *Examiner* in 1850. The anonymous author of the article wrote that 'Dr Latham's classification is based upon a combination of the physiological and psychological peculiarities of the different tribes and families of man'.[99] The reviewer added, 'we have only to add to the expression of our conviction that Dr Latham's book is destined to form an epoch in ethnological science'.[100] In the same year the *Prospective Review* wrote that '[Latham's] ... perpetual effort is after scientific and even mathematical form, as well as scientific accuracy. A sharp brevity of sentence, and a systematic arrangement of details, is perhaps natural to him, even when he does not aim at it'.[101] The reviewer continued by stating that Latham's style 'tends towards the perfection of the science; if it be not always altogether in conformity with English taste'.[102] Interestingly, however, the *Prospective Review* ended their assessment of Latham's work by comparing the *Varieties of Man* with Prichard's *Researches* and argued that

> Of the condensed learning contained in the book before us, we have scarcely given the reader an idea; but we may add, the reasonings are almost entirely those of literature, not of physiology, and although it is a worthy addition to Prichard, it is essentially supplementary, and does not attempt to work out great principles from the foundation.[103]

Thus they did not view the methods outlined in the *Varieties of Man* as a new approach that would carry the discipline of ethnology forward. Instead, they positioned it as an extension that built upon Prichard's sound principles. The *Prospective Review* maintained that Prichard's ethnological framework was the hegemonic discourse on the study of human diversity. However, it is striking that the *Prospective Review* framed Latham's book as primarily a literary study, when so much of his discussion was on anatomical topics.

In 1851 the *Athenaeum* also reviewed Latham's book and compared it to Prichard's *Natural History of Man* (1843), which was a condensed version of the *Researches*.[104] The reviewer believed that Prichard's methodological framework was the impetus behind many of the ethnological works released in Britain in the 1840s and 1850s. For example, they wrote that 'The work of Dr. Prichard on the "Natural History of Man" is not only a manual for the student of ethnography, but a favourite book with general readers'.[105] They continued by stating that 'His labours gave an impulse to inquire, – new men came into the field, – more accurate observations were made and recorded in every part of the world, – and in a few years ... the bounds of knowledge [were] greatly extended'.[106] This review from the *Athenaeum* is significant for two reasons. Firstly, it shows that Prichard's books – such as the *Natural History of Man* – were used as instructive manuals, and as such taught researchers and informants how to record a higher quality of ethnographic observations. Secondly, it reveals that his work motivated other ethnologists to further refine the research field's methods. Books such as *Varieties of Man* can be positioned as extensions of Prichard's ethnology. The *Athenaeum* also noted that with Prichard's passing in 1848, the discipline continued to transform because figures such as Latham were carrying the science forward. In particular, the journal recognized that Latham was able to further broaden the scope of the research field by using the latest ethnographic material available in Britain. They wrote, 'Dr. Latham indicates briefly the new sources of information which have been opened up, and the use which he has made of them in his present work'.[107] Overall it was a positive review of Latham's *Varieties of Man*, and it demonstrates that readers used large ethnological books to enhance and refine their ethnographic observational techniques.

In Charles Malcolm's 1851 presidential address delivered before members of the ESL, he discussed the publication of Latham's book. Malcolm stated that the society was 'justly proud' of Latham and that 'no one in this or any country ... approaches near him in that particular philological research with which the volume [*Varieties of Man*] largely enriched'.[108] However, Malcolm maintained that

> [Latham] has by no means neglected the geographical, the anatomical, the historical, and other circumstances which it is indispensable to take into consideration in an attempt to arrive at any thing like a thorough and accurate knowledge of the many difficult points which a systematic work on Ethnology must necessarily include.[109]

Malcolm argued that Latham's *Varieties of Man* was an important contribution to ethnology and ethnography because it clearly outlined the important features (physical and cultural) to consider when observing the various races of the world. Accordingly, Malcolm believed that Latham drew upon as many sources as were available to substantiate his suppositions. He ended his evaluation of Latham's work by stating that 'It is quite impossible to offer an epitome of such a book, which will doubtless be regarded as indispensable to the future cultivators of our science'.[110]

Knox's *Races of Man* met with a very different response throughout the periodical press, and for the most part his methods and theories were denounced.[111] Many reviews emphasized his disrespect for the British scientific and medical community, and in most cases journals positioned his strong opinions as the ideas of an embittered 'madman' unsubstantiated by empirical evidence. In 1850 the *Gentleman's Magazine* wrote that Knox 'repels us from a consideration of the great points of his work both by his abuse of all men and things which come in his way, and by the dogmatism of his unsupported assertions'.[112] The reviewer compared Knox's work to that of Prichard's, stating,

> We have lately seen our author's subject investigated by Prichard with laborious patience and acuteness, and in the true philosophic temper. How different is it in the hands of the present writer [Knox]. He rushes to the consideration of certain points of the great inquiry with a mind full of violent prejudices and in a rash presumptuous headlong spirit- overturning everything that stands in his way.[113]

This was a potent criticism because it was de-legitimized Knox's work as a politically motivated rant, while at the same time reinforcing the well-established and popular framework of Prichard and – by association – those researchers who built upon his ethnological model, such as Latham.

The *Gentleman's Magazine* was not the only journal to find Knox's attack on science and medicine overpowering and aggressive. In 1850 the *Literary Gazette* stated that 'a wild vein of political animosity runs through ... [Knox's] treatment of every subject, more in the style of the triton among the minnows of some country Athenaeum, than of a man who really must know something of the subject, and who possesses real merit for his account'.[114] In accordance with the way in which the *Gentleman's Magazine* challenged Knox's abilities as a scientific practitioner, the *Literary Gazette* argued that *Races of Man* was an 'amusing book in many respects'[115] because many of its claims were unfounded, with no evidence to support its suppositions. The reviewer also stated that they pitied 'the poor beginner into whose hands [Knox's] book fell'.[116] This was further significant because it identified a central problem with Knox's book: a confused structure and presentation. This was partially due to the fact that Knox published most of his writings in haste because of his desperate need for money.

As such, the *Literary Gazette* wrote that *Races of Man* was a difficult text for researchers and informants to use because it was hard for readers to navigate and locate important passages.

In the same year *Bentley's Miscellany* published an account of both Latham's *Varieties of Man* and Knox's *Races of Man*. The reviewer began by criticizing Knox's work because it challenged Prichard's monogenism and use of biblical theories and historical data. They stated that '[Knox] denies all that the Bible asserts as to the origin of man and the beginning of time, and calls chronology, sacred or profane, worthless, "and all human history as conjecture, pretension, error and obscurity"'.[117] This was a piercing critique of Knox's work. Moreover, *Bentley's Miscellany* found little of value in Knox's writings and ended its assessment by stating, 'we must close our notice of this offensive book'.[118] However, the reviewer managed to add one additional point by comparing Knox's ethnology to that of Latham: 'The work of Dr. Latham is of a totally opposite description; it is a scientific work, presenting a succession of new and most interesting facts throughout the whole of its pages, facts clearly stated and most methodologically arranged'.[119] Thus Latham's book was positioned as a more accessible resource for readers because of its clarity of exposition and organization.

Comparisons between Latham and Knox were present in other reviews. In 1851 a journalist for *Fraser's Magazine for Town and Country* wrote, 'Every page of Dr. Latham's ... [*Varieties of Man*] ... is marked by the impress of the accomplished scholar and linguist, the profound critic, and the subtle analyst'.[120] In contrast, the reviewer argued that 'Dr. Knox writes as a man who has a standing quarrel with all the world'.[121] It went on to say that 'Dr. Knox ... cannot take pen in hand without straightway getting into a passion, and placing himself, towards all who have the audacity to entertain opinions different from his own, in an attitude most unfavourable to the perception and discovery of truth'.[122] The reviewer from *Fraser's Magazine* stated that Knox explained human variation through the 'application of the transcendental theories ... for which our author betrays a very exaggerated and misplaced partiality'.[123] Thus readers did not readily accept Knox's competing ethnological model, and many of the reviews highlighted readers' scepticism towards his paradigm.

In 1852 Knox received a particularly scathing review from the *Athenaeum* that accused him of tainting the minds of youth with his attack on Prichard's ethnology: 'If he lectured as he writes, it may account for his popularity amongst young men entering on their professional career; since it is flattering to the vanity of the young to be told how their fathers have been led by the nose by great names, and what a rare opportunity they have of distinguishing themselves by the adoption of ideas altogether new in the world'.[124] This was significant because by the end of the 1850s the younger generation of practitioners such as Hunt and Blake later embraced Knox's alternative ethnological model. However, the *Ath-*

*enaeum* stressed that a more experienced reader would recognize the great 'flaws' in Knox's work and understand it to be nothing more than propaganda attempting to challenge the established authority of the leading members of the British scientific and medical community: 'The mature man finds out that he has been deceived: – and his warning to the rising generation may perhaps account for the fact of our meeting Dr. Knox so often now in literature'.[125] One of the main reasons that Knox was cited so frequently was because his work was controversial.

The reviewer went on to attack Knox's personal character, stating that 'Dr. Knox, in fact writes as a disappointed man; and so much that he says is evidently dictated by this spirit of disappointment, that his works can in no way be read as guides on the subjects which he professes to teach'.[126] The article was referring to the fact that Knox's career fell apart because of his connection to the Burke and Hare murders. It stated that 'if Dr. Knox would calmly reflect on his own career and productions, he would feel that he has lived in a house of a kind of glass that would not bear the return of such very large stones as he has been freely throwing at the houses of others'.[127] To sum up, looking at the ways in which readers interpreted the works of figures such as Latham and Knox is important as it represents a central component of any observational practice. It demonstrates the strengths and weaknesses of the works, and some of the ways in which practitioners used specialized literature in their research programmes.

## Conclusion: Towards the 'Anthropological Schism'

By the end of the 1850s there was a growing divide within the British ethnological community. On the one hand, figures such as Latham continued to transform and refine the observational practices of earlier researchers such as Blumenbach, Prichard and Lawrence. In particular, Latham built upon Blumenbach's taxonomical system for organizing human varieties into groupings based on physical characteristics and promoted a Prichardian monogenetic theory for the common origin of humans. At the same time, a separate group of practitioners led by Knox argued in favour of a polygenetic explanation for human origins and based their suppositions on racially deterministic theories influenced by French and German comparative anatomy and transcendentalism.

Reflecting on the closing years of the 1850s, James Hunt stated in his 1863 'Introductory Address to the Study of Anthropology' that 'Ethnology as now understood, has quite outgrown the narrow basis on which it was started. We must, therefore, enlarge and deepen our foundations; collect a range of facts, and extend our sphere of observations'.[128] Hunt continued by noting that 'An attempt has been made to divide all ethnologists into two parties, monogenists and polygenists: and each party is supposed to be bound to support the side to which they may be espoused'.[129] Hunt's evaluation of the state of ethnology in Britain at the start of the 1860s marked the beginning of an anthropological schism. The 1860s witnessed more theoretical and practical transformations, and there were many changes occurring within the two so-called conflicting approaches to the sciences of Man. It was only after the death of James Hunt (the most vocal opponent of Prichardian ethnology) in 1869 that practitioners from both camps negotiated a new anthropological synthesis.

# 4 THE BATTLE FOR MANKIND: JAMES HUNT, THOMAS HUXLEY AND THE EMERGENCE OF BRITISH ANTHROPOLOGY

> In a society like ours we want thinkers as well as observers. We should give every encouragement to [an] accurate reasoner, as it is to him we must look for the laws which can be deduced from our illustrations and accumulated facts.[1]
>
> James Hunt, 1863

In 1863 the physician and speech therapist James Hunt delivered his first presidential address before members of the newly formed Anthropological Society of London (ASL). In his speech, Hunt outlined an extensive practical framework for collecting ethnographic data, analysing the material and displaying its results. For Hunt – and other leading members at the ASL – the key to reforming the science lay in the observational practices researchers utilized in their studies. This was nothing new for the nascent discipline, as James Cowles Prichard, William Lawrence and Robert Gordon Latham continually sought to refine the observational techniques of ethnologists during the first half of the nineteenth century. However, as the 1850s came to a close, there was a growing divide within the British ethnological community. On the one hand, there were those researchers committed to a Prichardian observational model. Using a combination of anatomical and linguistic data, Prichard attempted to trace the common origin of humans by emphasizing similarities among the races. Conversely, there were those ethnologists interested in a Knoxian observational model that argued in favour of the multiple origins of races and promoted a methodology that prioritized inferences based on French and German comparative anatomy and transcendentalism.[2]

This chapter seeks to enlarge our understanding of ethnological and anthropological observational practices during the 1860s. Traditionally, historians and anthropologists who have examined the sciences of Man during this period have described the disciplinary divide as the 'anthropological controversy'. These scholars have framed members of the ESL as liberal promoters of a racially sensitive and inclusive ethnology and described their anthropological counterparts at the ASL as racists, sexists and supporters of an elite science.[3] Moreover, these accounts have situated the schism between the two societies as a competition over practice, although George Stocking did note in his book *Victorian Anthropology* that 'The contents of their journals suggest a largely overlapping intellectual field'.[4]

I want to expand upon this characterization by comparing the observational practices of Hunt and the anatomist, physiologist and biologist Thomas Huxley. Hunt and Huxley are good examples for this study because they have been regularly positioned as representing divergent methodological models. In this chapter, however, I will provide an alternative perspective by showing that the controversy was a professional competition and connected to the vocational strategies of both figures. By examining and comparing in detail the observational practices of Hunt and Huxley in turn, I will emphasize the commonalities between the two practitioners. Both of these researchers saw transformations in ethnographic collecting methods and the refinement of analytical techniques as key to reforming the discipline. They were at odds with each other because they were vying for control of the science. Subsequently, I want to explore the value for historians of taking seriously Hunt's and Huxley's self-descriptions as scientific reformers. Each believed he was strengthening anthropology and ethnology by emphasizing their best aspects and dispensing with their weakest.

Traditionally, the historiography has argued that Hunt's reason for leaving the ESL and starting up a rival society had to do with his sexism and racism. While these studies are suggestive, I want to re-examine both the woman and race questions in turn and connect them to Hunt's scientific reforms. By doing so, I want to suggest that behind Hunt's contentious views on extra-Europeans and the sexes was a larger motivating ambition to reform the observational practices of researchers. It is difficult to cultivate an intellectual sympathy for a figure such as Hunt because of his extreme views. However, he was influential in this period, and there is much to be gained by examining the factors underpinning his extremism. The participation of women in learned societies, monogenism, abolitionism and visual representations of human varieties all affected Hunt's vision of what the sciences of Man should comprise. I will also look at Huxley's ethnological reforms and link them to his views on women and race. As with Hunt, Huxley recognized the impact both these issues had on the intellectual scope of the discipline, as well as Victorian society as a whole.[5]

## Learning to Observe: Hunt and Huxley

> ... the most important new member [within the community] was a young man named James Hunt, who in a paradoxical and antithetical way was one of the most influential figures in English anthropology.[6]
>
> George Stocking, 1971

### *James Hunt: Physician, Speech Therapist and Scientific Reformer*

Historians have paid little attention to Hunt's early career, and it is important to note that before he became a figurehead for race discourse and physical anthropology, Hunt had a successful medical practice.[7] He built his career on a system for treating speech impediments developed by his father, Thomas Hunt (1802–51). Thomas was a speech therapist who developed an innovative technique for curing stammering and stuttering without recourse to surgery or the mutilation of patients' throats.[8] It was not long before Thomas Hunt's patient base grew. James took an interest in his father's work and learned the system. However, James improved upon his father's method by gaining a sound knowledge of both anatomy and physiology by studying medicine at Cambridge and the University of Giessen. While living in Germany, he studied French and German transcendentalism and comparative anatomy. Both featured prominently in his later anthropological works. Upon returning to Britain, he took over his father's practice after Thomas's death in 1851.[9] Hunt earned a substantial amount of money from speech therapy and even treated well-connected figures such as the novelist Charles Kingsley (1819–75). This was significant because it helped him network with some of the more influential people in Victorian London. When Hunt eventually founded the ASL, he invested much of his own money into the society. As one of the anthropological community's largest patrons, he could direct the activities and methodologies of researchers.[10]

Hunt's insistence during the 1860s that anthropology should be based on anatomical and physiological evidence derived from his medical background.[11] He was initially drawn to ethnology because of his interest in medicine and his concentration on speech and language. This was nothing new for the discipline. Prichard, Latham, King and Hodgkin incorporated their anatomical and physiological training into their studies. Hunt followed in this tradition, and he joined the ESL in 1856. As a young and ambitious researcher, he was very active at the society and consequently his status rose quickly. He was elected joint honorary secretary in 1859 and held this post until his resignation in 1862.

Following Prichard's death in 1848, the ESL struggled to find a new figurehead to lead the discipline. Hunt recognized that there was an opportunity for him to position himself as a scientific reformer, and he set out to establish and promote what he believed to be a sound disciplinary foundation built upon rigorous theoretical and methodological principles. His growing influence within the society was visible from an early stage. Hunt played a leading role in persuading the ESL to readmit Robert Knox into the society in 1858. This was significant because during the early 1850s Knox's contentious views on race were a continual problem for members such as the Quaker monogenists and abolitionists Hodgkin and Henry Christy. In 1855, after several major disagreements at meetings, the executive council had barred Knox from the society.[12]

Hunt was a keen promoter of ethnology and was busy networking with other scientific societies. He recognized the importance of communicating the activities of ethnologists to the larger scientific community, and in 1861 he published his first report on the proceedings at the BAAS. After the formation of the ASL in 1863, coverage of the activities of the ethnological subsection at the BAAS was a central concern for Hunt.[13] However, it was already clear to him in 1861 that the science should promote its research to a larger audience. In his report, Hunt 'expressed a hope [that] at the next meeting of the British Association at Cambridge, the Fellows of [the Ethnological Society of London] would attend in larger numbers; and that they should in the mean time [*sic*], use their best energies to induce both travellers and thinkers to send contributions which should be worthy of the science of Ethnology'.[14] That he asked members of the ethnological community to promote the science to other researchers and colonial informants potentially interested in the study of races demonstrates Hunt's desire to expand ethnological literature in Britain and its empire. It also shows that Hunt was actively involved in the initiatives of other ethnologists – such as King and Hodgkin – who during the 1840s and 1850s wanted to build an ethnographic exchange network with informants living throughout the world.[15] Thus, before his exit from the ESL in 1862, Hunt was actively engaged in transforming British ethnology.

Hunt's commitment to reforming ethnology soon began to wane, and one of his earliest worries regarding the discipline's development had to do with its position at the BAAS. In the 1850s ethnology shared section E with geography at the BAAS. Hunt was particularly concerned about the lack of a section devoted exclusively to ethnology, and he pointed out that 'Altogether there were twenty-five papers read before the Section. Many of them, he thought were foreign to ... ethnology'.[16] Hunt believed that without an autonomous section for ethnologists to present and discuss their research at the BAAS, the expansion of the science would be stunted. Furthermore, he became convinced that speculative theory, philanthropy and Christian morality and sentimentality dominated ethnology at the BAAS and within the ESL.[17]

It was evident by the start of the 1860s that Hunt was frustrated with what he viewed as the central methodological focus of ethnology: an overemphasis on non-physical data such as linguistic evidence, and an over-reliance on biblical explanations for the origin of humans. Moreover, Hunt had become interested in comparative anatomy and transcendentalism during his time living on the continent and through his close relationship with his mentor Knox. For those reasons, Hunt believed that the discipline lacked sufficient observable data – such as anatomical and physiological evidence – to substantiate its theories. Soon he and his supporters broke away from the ESL and formed a new society dedicated to promoting the sciences of Man according to Hunt's anthropological vision. Their polygenetic views and focus on comparative anatomy and transcendentalism were incompatible with those of ethnological monogenists, who were increasingly interested in Darwinian evolutionary theory. This was a further hindrance for Hunt because among those new members who sought to redefine ethnology along Darwinian lines were some of the more influential scientific promoters of the day, including the scientific naturalists Alfred Russel Wallace (1823–1913), John Lubbock (1834–1913) and Huxley. Each would pose a threat to Hunt's authority within the ESL.[18] In addition, in the early 1860s Hunt and his associates organized a private collective known as the 'Cannibal Club', which acted as a counterpart to the scientific naturalists' exclusive dining group known as the 'X Club'. At these dinners 'the Cannibals' would discuss their research with each other and develop strategies for promoting their scientific vision.[19]

## *Thomas Huxley: Anatomist, Physiologist, Biologist and Scientific Reformer*

> I have a reputation (of which, I fear, you will not approve) of being a great heretic and a savage controversialist always in rows.[20]
>
> Thomas Huxley, 1864

Huxley took a slightly different path into ethnology than Hunt. Nevertheless, there are several similarities between their social backgrounds, training and vocational strategies. What will become clear is how all three of these aspects in their lives shaped their observational practices. As with Hunt, Huxley came from a middling-sort background. His father, George Huxley (1780–1853), was a schoolmaster in Ealing Middlesex before moving his family to Coventry in 1835 where he acquired a job as a banker.[21] Early on, Thomas Huxley took an interest in anatomy and physiology, and under the tutelage of his medically trained brother-in-law John Charles Cooke, he decided to enter the medical profession. Many ethnologists during the first part of the nineteenth century possessed some training in medicine, and anatomical and physiological studies formed an important component of the discipline's practices. Huxley's training in anatomy and physiology would become a significant part of his ethnological methodology in the 1860s, and – as with Hunt – he argued that all studies on human diversity must be supported with directly observable data.[22]

Compared to many of his contemporaries, Huxley's medical training was much more hands-on. In preparation for medical school, he was apprenticed under the surgeon Thomas Chandler in the East London dockland area. In addition, instead of learning medicine at a medical institution such as the one at Cambridge – where there was less opportunity to directly engage with bodies – Huxley studied at the newly formed Charing Cross Hospital in London, where he gained first-hand experience treating patients and working with human cadavers.[23] One of his anatomy teachers was Knox, who was Hunt's primary mentor on human variation topics. Huxley also possessed direct experience observing extra-Europeans *in situ*, because his first job after completing his medical schooling was as an assistant-surgeon on board the HMS *Rattlesnake* (1846–50). This voyage was an important developmental period for Huxley because it provided him with first-hand knowledge of the indigenous populations of the South Pacific in their own environment. Thus, unlike figures such as Prichard from the earlier part of the nineteenth century, Huxley could combine his own travel observations with information collected by informants when writing his ethnological works.[24]

Huxley was also young and ambitious. However, unlike Hunt, he did not inherit a family practice on which to build his career. When he returned from his voyage to the South Pacific he struggled to find employment, and it was not until the naturalist Edward Forbes vacated his post at the Royal School of Mines in 1855 that Huxley finally obtained a permanent position. Once he secured his post, Huxley's rise to prominence did not take long.[25] In the same year that he replaced Forbes at the RSM, he also became a naturalist for the Geological Survey and the Fullerian Professor at the Royal Institution. In addition, Huxley was appointed as the Hunterian Professor at the Royal College of Surgeons in 1863 and joined the ESL, becoming its president by the end of 1869.[26]

Huxley was concerned about the state of British ethnology when he first joined the ESL, and one of his earliest contributions to the discipline was in promoting the science to a larger audience. Leonard Huxley noted in the *Life and Letters of Thomas Henry Huxley* that his father's chief interest during the 1860s lay in ethnology, and this is further evidenced by his father's lecturing and publishing activities.[27] During the 1860s Huxley delivered several lectures on ethnology at the Royal College of Surgeons and at working men's clubs. He also wrote two of his most important works on ethnological subjects during the 1860s: *Evidence as to Man's Place in Nature* (1863) and his article 'On the Methods and Results of Ethnology' (1865).[28]

We can see several parallels between Hunt and Huxley. Both trained in medicine and were interested in promoting and reforming the sciences of Man. For each of our figures, the key to the discipline's transformation lay in the observational practices researchers utilized in their studies. However, they were both vying for the leadership of the research field, and as a result they soon became arch-rivals. Some historians such as Edward Caudill and James Elwick have suggested that Huxley's strategy for reforming most natural sciences followed a similar process. First he would introduce Darwinian evolutionary theory into the discipline.[29] This was fairly straightforward for ethnology because many of its practitioners already argued in favour of common descent, and topics relating to species and varieties were at the forefront of ethnological discussions.[30] The next step for Huxley was to make the science ideologically neutral, separating it from both religious and political debates.[31] It was a highly effective strategy for achieving an influential social status, because Huxley and the other scientific naturalists such as John Tyndall (1820–93) could emphasize the important and impartial perspective the sciences had on a broad range of issues.[32] The third step for Huxley was to villainize researchers in the discipline who did not adopt an evolutionary stance and scientific neutrality. This was a strategy that Hunt and Huxley shared. Both targeted established scholars who clung to older ideologies that they viewed as outdated. Hunt argued that monogenists such as Hodgkin and Christy overcommitted to biblical explanations for the origin of Man. Within the biological sciences, Huxley targeted Richard Owen (1804–92), framing him as an antiquated naturalist and representative of early nineteenth-century practices.[33]

Huxley, however, altered his vocational strategy when he challenged for the leadership of the sciences of Man. One of the reasons he struggled to establish himself within the scientific community when he returned from the *Rattlesnake* voyage was because the older guard of naturalists in Britain were striving to maintain their authority over the various research fields. Consequently, once Huxley finally became secure within the community, he was ruthless in maintaining his dominance. However, Hunt's ambition for control of the sciences of Man posed a new problem. As with Huxley, and the other scientific naturalists such as Tyndall and Lubbock, Hunt was a young and charismatic speaker. He saw the promotion of the discipline to a larger audience as key to its transformation. Moreover, he wanted to free the sciences from religion and push out older scholars who grasped to archaic practices. Evelleen Richards argued that

> [Hunt's] ... was a hegemonic struggle with Huxley and the Darwinians. The object was to define the ideological role of anthropology in Victorian Society. In a period when traditional theological modes of explanation were giving way before a secular redefinition of the world, the Anthropologicals and the Darwinians offered two competing versions of a legitimizing scientific naturalism.[34]

In view of their commonalities, Huxley was limited in his tactical approach to discrediting Hunt. One of the few significant points of contention for our two figures was that Hunt supported slavery and the Confederates during the American Civil War (1861–5), while Huxley remained primarily neutral on the subject. Hunt's over-commitment to proving that Africans and Europeans were fundamentally different at a physical level was partially responsible for his loss of authority.[35] Another key factor was that Hunt routinely broke many of the codes of gentlemanly behaviour within Victorian society by attacking the character of many of his rivals in his self-financed periodical the *Anthropological Review* (*AR*).[36] There were noticeable similarities between Hunt's and Huxley's training and vocational strategies. These commonalities also manifested themselves in their observational practices. In particular, when we begin to examine in detail the various techniques employed by our two figures within their practical frameworks, we see even more overlap. Both Hunt and Huxley recognized the significant impact that the woman and race questions had on the intellectual and methodological scope of ethnology and anthropology.

## Women, Race and the Study of Man

> What social and political rights have women? What ought they be allowed, or not allowed, to do, be, and suffer? And, as involved in and underlying all these questions, how ought they to be educated?[37]
>
> Thomas Huxley, 1865

### *Women's Place within the Sciences of Man*

It is difficult to cultivate an intellectual sympathy for a figure such as Hunt because of his extreme views. However, in reassessing the causes behind the anthropological schism of the 1860s, it will be helpful to reconsider Hunt as he saw himself, not as a racist and sexist, but as a scientific reformer. There are several reasons that contributed to Hunt's resignation from the ESL in 1862. Richards argued that Hunt resigned from the ESL because of his annoyance with the President John Crawfurd (1783–1868), who decided to admit women to their society's meetings. George Stocking and Douglas Lorimer have put more emphasis on the issue of race. Although these arguments are suggestive, they oversimplify Hunt's motivations by overlooking the broader context of his dissent. Hunt's disapproval of women at learned societies and his views on race both connect to a couple of larger issues: observational practices and the circulation of scientific knowledge. Huxley also engaged in both of these debates, and by comparing our figures' views on these questions, it is clear that there is much overlap between their ideological and practical models.

Hunt revisited the woman question in his 1868 article 'On the Origin of the Anthropological Review and Its Connection with the Anthropological Society', in which he noted that 'one of the earliest matters of dispute was with respect to the admission of ladies to the meetings of the [Ethnological] Society'.[38] Hunt and his supporters argued that a woman's presence at ESL meetings would limit the ability of its members to speak openly about topics relating to human diversity. For Hunt, the woman question was an issue because it affected the direction he envisaged as the correct future path for the discipline. If ethnological research were to focus predominantly on cross-comparative philological and linguistic studies, the admittance of woman to the society was less problematic. However, Hunt wanted ethnology to substantiate its research claims by primarily using anatomical and physiological evidence, including discussions that were deemed inappropriate for Victorian women.

In his 1864 dedication to Paul Broca (1824–80) from the ASL's translated edition of Carl Vogt's *Lectures on Man*, Hunt discussed the significance of the woman question for the refinement of the science, and he argued that ethnology required 'the most free and serious discussion, especially on anatomical and physiological topics'.[39] However, if women were present at meetings, researchers would be unable to examine subjects such as male genitalia among different races. This would in turn affect the *Transactions of the Ethnological Society* because the submissions would reflect the discussions and papers presented at the society's meetings. Hunt wrote that 'The party under his leadership [Crawfurd] desired to place the Ethnological Society on a footing with the Royal Geographical Society, and to render its meetings fashionable and popular by the admission of ladies. You will, doubtless smile at a strange idea of admitting females to a discussion of all ethnological subjects'.[40] Accordingly, Hunt's disapproval of women attending ESL meetings was not simply because he believed they were incapable of grasping the scientific pronouncements of ethnology. Rather, Hunt framed his objection by arguing that their attendance at the meetings would impede the expansion of the science and would affect ethnological knowledge and practice. By restricting the topics that the fellows could discuss, ethnologists were stunting the intellectual scope of certain research areas. However, Hunt was a masterful propagandist and utilized the admittance of women at the society to support his exit.

In light of the cultural climate of the 1860s, it is possible that Hunt utilized the admittance of women at the ESL as a convenient pretext to support his exit from the society, and that this was part of his professional strategy. Given that other leading scientific figures such as Huxley had reservations about allowing women to participate in scientific forums, it is easily conceivable that Hunt exploited the popular opinion of the general scientific community as a reason to legitimize his departure. In doing so, Hunt and his supporters looked like defenders of traditional scientific practice, and they became the upholders of conventional science, which excluded women from participating. For Hunt, it was crucial to portray himself and his supporters as the conventionalists in this conflict. John Crawfurd (who was president of the ESL during the dispute) became the radical liberal because of his desire to admit women into science.

It is suggestive that Hunt rarely referred to the woman question during the years of 1863–8. It was not therefore his primary criticism of the ESL. He only revisited the issue in 1868 because of the attacks he suffered from the larger scientific community at the height of the anthropological controversy.[41] Interestingly, it was possible for wealthy women to support anthropological research monetarily. One of the first motions carried out by the ASL on 5 August 1863 was to allow women to invest into their society. Richards recounted that '"Ladies" might become financial members of the Anthropological Society, but "shall on no occasion whatever be allowed to attend any of the meetings of the Society". Their cash was welcome, but not their persons'.[42]

Huxley's position regarding the woman question largely overlapped with Hunt's perspective. Huxley argued in his famous article 'Emancipation – Black and White' (1865) that woman had the right to an education. He wrote that educators should 'Recognize the fact that [both sexes] ... share the senses, perceptions, feelings, reasoning powers, [and] emotions ... so that whatever argument justifies a given education for all boys justifies its application to girls as well'.[43] As with Hunt, for Huxley the problem was not that women were incapable of comprehending the scientific pronouncements of ethnology. Rather, it was the nature of the subject matter that was the issue. Huxley first realized the significant impact women's attendance at meetings had on the development of ethnology after an exchange of letters with Wallace in 1864. Huxley wrote in an epistle to his friend that he saw no value in having two societies based in London devoted to the scientific study of Man, and he believed that Hunt's decision to break away from the ESL was unjustified. However, Wallace responded to this criticism by emphasizing the importance of holding meetings with just male members, and he wrote that

> I cannot agree with you that 'there was not the slightest reason for [the Anthropological Society's] existence'. It seems to me that its establishment is a good protest against the absurdity of making the Ethnological a ladies' Society. Consequently many important and interesting subjects cannot possibly be discussed there ... the Anthrop. is the only place where these can be discussed.[44]

Wallace's defence of the ASL had an effect on Huxley, and one of his first moves as president of the ESL was to deal with the issue of women at society meetings. As with Hunt, Huxley concluded that women attending lectures impeded the intellectual and methodological scope of the discipline as a whole.

The admittance of women to the society was a key concern for the old guard of ethnologists at the ESL. Figures such as Crawfurd, Christy and Hodgkin championed an inclusive science open to both genders. However, by the late 1860s the scientific naturalists such as Lubbock and Edward Burnett Tylor (1832–1917) were in the ascendancy within the ethnological community, and by 1868 Crawfurd, Christy and Hodgkin had passed away. Thus it was much easier for Huxley to reverse the council's original decision to admit women. In 1869, against the protests of female members such as the early feminist Eliza Lynn Linton (1822–98), the ESL executive council, led by Huxley, announced that the society would host two types of meetings for its members.

The first type was 'ordinary meetings' where 'Ladies will not be admitted. The subjects will be scientific, and their treatment such as will be calculated to promote close study of the subjects'.[45] In other words, discussions at these meetings could focus on all topics because it was unimpeded by the presence of women. This would put the ESL on an equal ground with the ASL. The second type was 'special meetings' where 'Ladies will be admitted. The subjects will be popular, calculated to secure the interest of the public'.[46] This was an astute move, because by having two types of meetings, Huxley could promote the close study of all ethnological subjects at the society, while at the same time maintain one of the long-established tenets of the earlier generation of ethnologists who wanted to open the discipline up to everyone. Moreover, 'special meetings' would help to promote the science to a broader audience without bombarding attendees with complicated terminologies aimed at more specialized practitioners. As we can see, both Hunt and Huxley had similar ideas about the place of women in science, and each saw the effect their presence at society meetings had on the discipline's practices. If the sciences of Man were to continue to reform and expand its methodological frameworks, it was important not to limit the scope of its research domain.

## *The Race Question and Debates about Representation*

Another cause historians have attributed to Hunt's resignation from the ESL relates to his perspective on race. As with gender, Hunt's extreme racial views related to his programme to transform the research field, and knowledge and practice was at the centre of his reforms.[47] In 1861 the ESL assigned Hunt a leading role in reviving the society's defunct periodical, the *Journal of the Ethnological Society of London*.[48] While he was working on this task, a disagreement erupted in May 1863 between Hunt and several senior members of the ESL, including Hodgkin and Christy. The evidence is patchy, and the ESL recorded little detail of the disagreement. However, with the archival material available, it is still suggestive that observational practices played a central part in the debate. Evidently, the two sides disagreed about both the cost and the depiction of some lithographs from an article on the inhabitants of Sierra Leone, especially its portrayal of Africans.[49] The significance of this dispute is more understandable when one considers the larger debates occurring at the ESL between Hunt, the racial determinist with polygenetic sympathies, and Hodgkin and Christy, who were devout monogenists and supporters of abolitionism.

Hunt was increasingly interested in scientific discussions about human variation, and in several of his articles he propagated extreme views about race.[50] Hunt acknowledged that his racial theories were 'imbibed from the late Dr. Knox', who attributed everything to racial determinism.[51] In addition, Hunt believed that there was insufficient observable data available to substantiate any theory that attempted to explain the origin of humans. Instead researchers were to focus on the contemporary state of races. With this in mind, Hunt believed that Africans were most likely a separate species from Europeans because their physical forms were visibly different.[52] He argued in his famous treatise *On the Negro's Place in Nature* (1863) that

> It is generally taught that the Negro only differs from the European in the colour of his skin and the peculiarity of his hair, but such opinions are not supported by facts. The skin and hair are by no means the only characters which distinguish the Negro from the European, even physically; and the difference is greater, mentally and morally than the demonstrated physical difference.[53]

Because Hunt believed that Africans possessed inferior cognitive abilities than those of Europeans, he also argued that their capacity to become civilized was limited. He wrote, 'The assertion that the negro only requires an opportunity for becoming civilized, is disproved by history'.[54]

In light of Hunt's extreme racial views and his ambition to reform the observational practices of researchers within the sciences of Man, we should reinterpret the debate regarding the use of the Sierra Leone lithographs for the society's periodical as both a conflict about scientific accuracy and a platform for him to express his contentious views on race. The clash between Hunt and the senior members at the ESL – including Hodgkin and Christy – is further complicated by political motivations. Since 1787 Sierra Leone was a colony for the resettlement of free and emancipated Africans, and by the 1860s it was also a central location for abolitionists and humanitarians to demonstrate the 'unity of race'.[55] Despite claims by extremists such as Knox and the American physician Josiah Nott (1804–73) that Africans were unable to reach the same level of civilization as white Europeans, the African colony at Sierra Leone was stable.[56] For Hunt, the publication of the article on the people of Sierra Leone was problematic because it communicated to the readership a depiction of Africans which varied from his own perspective and which he believed was incorrect.

The article 'Sketches of the Colony of Sierra Leone and its Inhabitants' that accompanied the lithographs was politically motivated. The author, Robert Clarke, was a colonial administrator in Sierra Leone. He discussed the inhabitants of the region, its climate and economy, and without openly stating his political orientation, Clarke wanted to support the notion that Africans could achieve the same level of civilization as Europeans. For example, he emphasized the moral strength of the African character in resisting vices such as alcoholism and public drunkenness. In Britain, drunkenness was utilized by social commentators to illustrate the moral collapse of British society.[57] Likewise, Clarke sought to demonstrate the stability of the Sierra Leone colony and its people by asserting that 'The manners and customs of several of the negro nations might, however, in many points of resemblance, bear a comparison with those of the European'.[58] Thus there were obvious arguments for the potential of racial parity within the text.

Hunt's conception of Africans clashed with Clarke's depiction of the Sierra Leoneans. In his treatise published in the same year, *On the Negro's Place in Nature*, he argued that

> It is alleged that the Negro only requires early education to be equal to the European; but all experiments of this kind have proved that such is not the fact. With the negro, as with some other races of man, it has been found that the children are precocious: but that no advance in education can be made after they arrive at the age of maturity, they still continue, mentally [to be] children.[59]

From Hunt's perspective, Clarke's account of Africans was factually incorrect, and the images that accompanied the text further misrepresented the population.

The images reproduced in the *Transactions of the Ethnological Society of London* (see Figures 4.1–4.3) portrayed thirteen Sierra Leoneans drawn by Mrs Clarke, who was Robert Clarke's wife.[60] Unlike many contemporary illustrations of Africans, which depicted them as almost inhuman, Mrs. Clarke's drawings presented the reader with sketches of Africans with refined features and European style clothing.[61] Hodgkin and Christy probably supported the publication of these images because they demonstrated the civility of the Africans living in the free colony of Sierra Leone. It served to reinforce notions that Africans had the potential to become equals to Europeans. As a polygenist and staunch supporter of slavery, Hunt did not approve of the publication of these images. It presented a view which conflicted with his own ideas about race and demonstrated that Africans were not 'savages' but could develop into members of civilized societies.

Figure 4.1: A lithograph of a Sierra Leonean woman, sketched by Mrs Clarke. From R. Clarke, 'Sketches of the Colony of Sierra Leone and its Inhabitants', *Transactions of the Ethnological Society of London*, 2 (1863), pp. 320–63, unnumbered insert between pp. 347 and 348. Reproduced from the author's private collection.

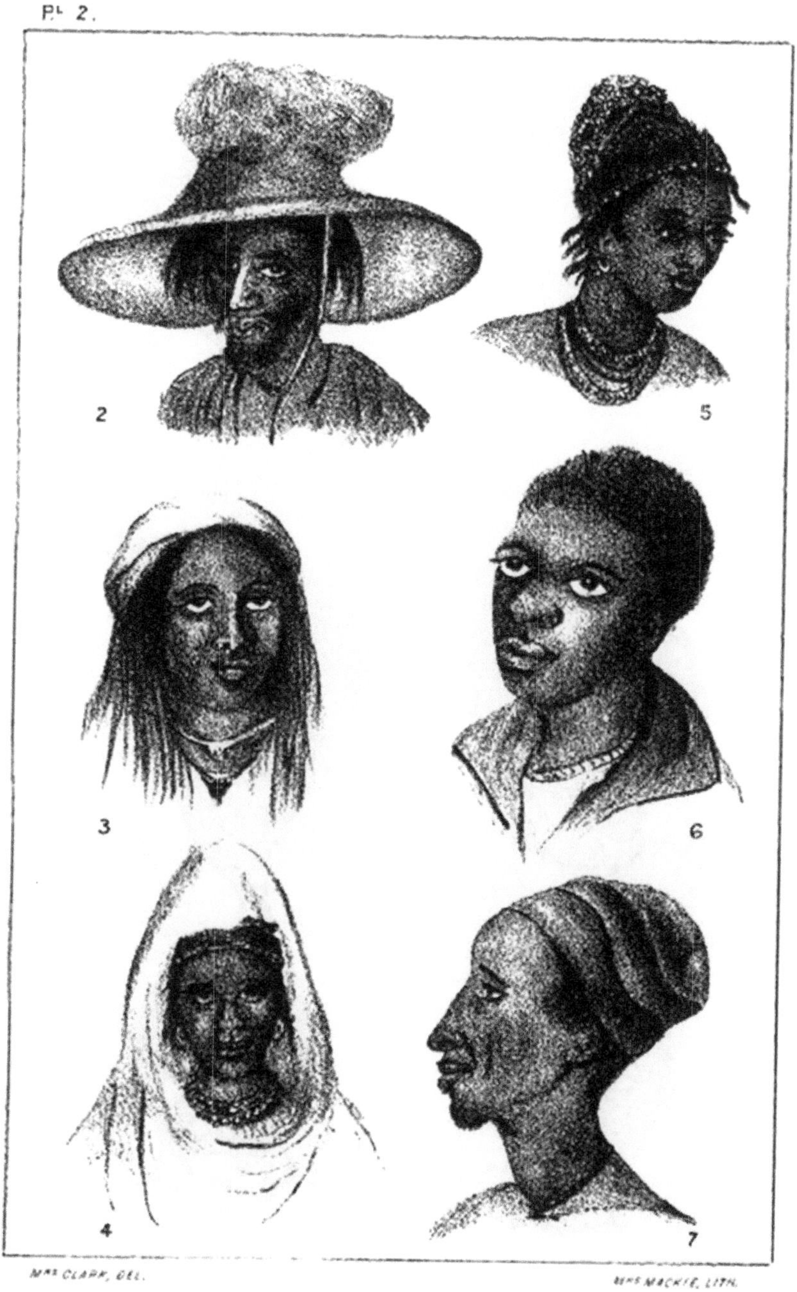


Figure 4.2: A lithograph of several Sierra Leoneans, sketched by Mrs Clarke. From R. Clarke, 'Sketches of the Colony of Sierra Leone and its Inhabitants', *Transactions of the Ethnological Society of London*, 2 (1863), pp. 320–63, unnumbered insert after p. 352. Reproduced from the author's private collection.

Figure 4.3: A second lithograph of several Sierra Leoneans, sketched by Mrs Clarke. From R. Clarke, 'Sketches of the Colony of Sierra Leone and its Inhabitants', *Transactions of the Ethnological Society of London*, 2 (1863), pp. 320–63, unnumbered insert after p. 352. Reproduced from the author's private collection.

Robert Clarke carefully positioned the legitimacy of his wife's drawings in accurately depicting the physical characteristics of her subjects: 'Before submitting these lithographs to the public, I may be permitted to state that, although Mrs. Clarke is no artist, nevertheless the original drawings are accurate and truthful likenesses. And so much is this the case, that the portraits were immediately recognized by those who know the different individuals of the races represented'.[62] In asserting the accuracy of his wife's illustrations, Clarke transformed them from artistic representations of the inhabitants of Sierra Leone into credible ethnographic images for scientific investigations.

The ESL executive council met to consider Hunt's case for rejecting the images. Unsurprisingly, Hunt failed to convince the council, which included Hodgkin and Christy. The details of this debate are vague, and the ESL meetings record nothing of the dispute until 5 May 1863, when it was noted that there was some 'differences of opinions having arisen in relation to the woodcut design to illustrate the paper on the Natives of Sierra Leone by Mrs. Clarke, it was resolved that the arrangement of the question be referred to the Honorary Secretary Mr Wright'.[63] The ESL members elaborated no further on the issue and clearly considered it a closed matter. However, the conflict must have frustrated Hunt because after this meeting he resigned from the ESL. His resignation is highly suggestive because the issues of knowledge and practice influenced Hunt's decision to leave.

In the preface to the *Transactions of the Ethnological Society of London* from 1863, the secretary of the ESL, Thomas Wright, discussed the images of the Sierra Leoneans. There was an opportunity to present the conflict with Hunt to the readership, but Wright carefully avoided the occasion. Instead, he wrote that the engravings were 'taken from a series of beautiful drawings of the natives of that colony'.[64] There was no mention of Hunt or the debate that ensued regarding the images' publication. Nor was there reference to the newly formed ASL anywhere in the 1863 volume. Wright noted that 'The publication of the present volume of the *Transactions* of the Ethnological Society has been retarded by a combination of circumstances which it was not possible to foresee or prevent, in the present instance, but will be provided against in future'.[65]

Wright did not connect these 'unforeseen circumstances' to Hunt's resignation. Certainly, it raises the question as to whether the ESL viewed Hunt's departure as controversial. The separation of the two societies lasted eight years. None of the volumes from the *Transactions of the Ethnological Society* throughout the years of 1863–9 (its entire run) discussed the anthropological schism. One reason for this omission may relate to issues of credibility and authority. If the ESL ignored the activities of the ASL, it was in essence disregarding its legitimacy as a scientific body and its claims to a specialized scientific knowledge. This is further evidenced by Huxley's refusal to recognize their society's scientific activities. For instance, in a letter to Charles Darwin on 2 July 1863, Huxley referred to the ASL as 'absurd', and wrote that he 'sent back [their] Diploma of Honorary Fellowship'.[66] In contrast, Hunt maintained a loose affiliation to the ESL, and the *Transactions* from 1863 listed him as an honorary member of the society.[67] This was not an aberration, because in many cases members of the ESL attended ASL meetings, and vice versa.[68] Wallace, for instance, was an active member of both societies. Thus the distinction between the two groups was difficult at times to see. It was not as straightforward as being either an anthropologist in support of Hunt or an ethnologist protesting against his initiatives.

Nevertheless, how one observed, analysed and represented human variation was a central concern for Hunt, and it is clear that the debates surrounding the use of the lithographs connected to ethnological discussions about knowledge and practice. It is evident that Hunt recognized the important role the *Transactions* played in shaping its readers' understanding of race. That said, within the ESL Hunt's racial views and focus on polygenetic theory would continue to be at odds with those of monogenists such as Hodgkin and Christy, as well as evolutionary-minded scientific naturalists such as Huxley, Lubbock and Tylor. Clearly, the result of these differences of opinion over the Sierra Leone lithographs in conjunction with the woman question influenced Hunt's decision to leave the ESL and form a separate society. If Hunt wanted to reform the sciences of Man and disseminate his vision of the discipline, it would not be at the ESL. Therefore Hunt formed the ASL in order to promote his version of the science with researchers who shared similar views.

Traditionally, the historiography has positioned Huxley as the liberal thinker, whose ideas blended easily with those of the older guard of ethnologists. However, his perspective on race had more in common with Hunt than it did with Hodgkin and Christy. Huxley was of two minds about the American Civil War. On the one hand, he believed that slavery was an outdated practice and an insufficient economic system for a country to maintain. On the other hand, his favourite sister Lizzie lived in Tennessee, and her husband was a surgeon for the Confederates. Thus Huxley had familial ties to Southern politics. In addition, unlike Hodgkin and Christy, Huxley was not sympathetic to the Africans' plight. In a letter to his sister Lizzie from 4 May 1864, he wrote,

> I have not the smallest sentimental sympathy with the negro; don't believe in him at all, in short. But it is clear to me that slavery means, for the white man, bad political economy; bad social morality; bad internal political organization, and a bad influence upon free labour and freedom all over the world. For the sake of the white man, therefore, for your children and grandchildren, directly, and for mine, indirectly, I wish to see this system ended.[69]

Huxley reiterated this point publicly in his 1865 article 'Emancipation – Black and White', where he stated that

> The doctrine of equal natural rights may be an illogical delusion; emancipation may convert the slave from a well-fed animal into a pauperized man; mankind may even have to do without cotton shirts; but all these evils must be faced, if the moral law, that no human being can arbitrarily dominate over another without grievous damage to his own nature, be, as many think, as readily demonstratable [*sic*] by experiment as any physical truth.[70]

Abolishing slavery for Huxley had little to do with the exploitation of Africans and the natural rights of all humans. Rather, it had to do with the moral consequences it had on Caucasians. Such a stance placed him much closer to Hunt and the anthropologicals than it did to monogenists such as Hodgkin and Christy. Moreover, Huxley argued that scientifically objective representations of human varieties was a primary concern for ethnology, and those depictions began with observable evidence based on anatomical and physiological descriptions.[71] Therefore, as with the woman question, Huxley's ideas about race were similar to Hunt's racial ideologies, and they clearly affected his writings on the sciences of Man. Unlike figures such as Prichard and Hodgkin, he was not striving to demonstrate that all humans were of one origin to support abolitionism. Instead, Huxley aimed to accurately depict the different races that inhabited the world according to the observable evidence available.[72]

## Observational Practices, Natural Selection and the Study of Human Diversity

> We recently heard discussions respecting Man's place in nature; but it seems to me that we err in grouping all the different races of Man under one generic name ... If we wish to make any advance in discussing such a subject, we must not speak of man generally, but must select one race or species, and draw our comparisons in this manner.[73]
>
> James Hunt, 1863

### *Hunt and the 'New' Science of Anthropology*

It is difficult to write about Hunt because – as with his mentor Knox – his correspondence and notebooks do not survive. Any examination of Hunt's research activities requires that the historian piece together evidence from several sources. However, Hunt was an important contributor to the development of anthropological observational practices, and an examination of his methods sheds new light on the history of the discipline and race studies during the 1860s. In 1863 Hunt defined anthropology's research scope in his 'Introductory Address on the Study of Anthropology'. He wrote, 'Anthropology is ... the science of the whole nature of Man. With such a meaning it will include nearly the whole circle of sciences. Biology, anatomy, chemistry, natural philosophy, and physiology must all furnish the anthropologist with materials from which he may make his deductions'.[74]

For Hunt, the scientific study of races had to base its deductions on directly observable evidence. He criticized studies that attempted to explain the origin of humans. He believed there was insufficient evidence available to substantiate any origin theory, and he wrote, 'What time has not been wasted in idle speculations, assumptions and theories respecting the history of Man!'[75] Such a stance was nothing new for the discipline. Hunt's mentor Knox argued throughout his career that direct observation was the cornerstone of any analysis on human variation, and he criticized figures such as Prichard, claiming they were committed to explaining human origins through non-empirical suppositions.[76]

Hunt's devotion to an empirically based science grounded primarily on anatomical and physiological data is evident throughout his collected works. In his 1867 article 'On Physio-Anthropology, its Aim and Method', he wrote, 'I shall further attempt to show that our future success in the establishment of a reliable physiology of mankind will chiefly depend on freeing our inquiries from metaphysical speculation, and entering on the field of direct observation'.[77] Hunt believed that the first step for any anthropological study was to collect as many observable 'facts' as possible before interpreting their meaning. He wrote, 'My immediate object is to give an opportunity for discussing most fully and carefully what is the right method of research respecting … mankind'.[78] He continued by arguing that

> This question of method I regard as so important that I desire on the present occasion to forego the discussion of controversial points in order that our sole attention may be given to its consideration … What have we been fighting against during the last four years? Simply to abandon all theological, metaphysical, and *à priori* assumptions, when unsupported by observed facts.[79]

The way in which researchers made sense of these 'facts' was equally important, and Hunt argued that 'facts are of little service to the progress of genuine science, unless they are used properly'.[80] Thus he argued that the best method of analysis when working with evidence was the Baconian method of induction, which had three main steps. Firstly, researchers were to collect material and describe the 'facts'. Secondly, they were to tabulate or classify the 'facts' into three categories: a) instances where a specific characteristic was observed, b) instances where the characteristic was absent, and c) instances where a variation of the observed characteristic was present. Thirdly, in light of what the tabulated material demonstrated, researchers were to draw conclusions based on the data and determine which phenomena (physical or cultural) were connected to it, and which phenomena were not.

Hunt claimed that the use of the Baconian method of induction for analysing material was a distinct practice for anthropological research, and this was part of his professional and disciplinary strategy for separating anthropology from ethnology. This approach to strengthening anthropology's research claims was built on a long-standing tradition within Britain. For instance, it was similar to how leading members of the Geological Society of London (f. 1807) utilized Baconian methods to substantiate their research claims in the earlier part of the nineteenth century.[81] Hunt argued that anthropologists' commitment to Baconian practices made their studies more credible. He wrote,

> It has been solely the application of this method which has given such weight to our deliberations and our deductions. Loyalty to facts with regard to ... anthropology brought us face to face with popular assumptions, and the contest has resulted in victory to those who used the right method. Having then seen the advantage of conducting our investigations ... according to the inductive method ...[82]

However, this was a rhetorical argument, and ethnologists such Prichard, Lawrence and Latham utilized a Baconian method of induction when analysing their data. The only difference was that Hunt explicitly identified the importance of it because he was attempting to legitimize his methodological approach as being a more refined observational practice. [83]

We see examples of Hunt's strategic use of the Baconian method throughout his writings. In his treatise *On the Negro's Place in Nature*, he criticized Blumenbach and Prichard for drawing conclusions about human varieties without properly tabulating their material and conducting extensive comparative studies.[84] Accordingly, one of Hunt's primary challenges to the Prichardian observational framework was that it did not provide enough quantifiable and numerical data for the different races of the world. Hunt claimed he was much more statistical in his descriptions. He wrote, 'The extremities of the Negro differ from other races more by proportion than by form: the arm usually reaches below the middle of the femur. The leg is on the whole longer, but is made to look short on account of the ankle being only between 1 ⅓ in. to 1 ½ in.'[85] Numerical data was often represented as a more empirical type of evidence, and polygenists routinely used this evidentiary tactic when attempting to discredit monogenists.[86]

Using this Baconian technique as a strategy to strengthen his research claims, Hunt also included tabulated data in his account *On the Negro's Place in Nature* (see Figure 4.4). Because he prioritized observations based on first-hand experience, he utilized material collected by the German physician and colonial informant Franz Ignaz Pruner (1808–82) to substantiate his claims. Pruner taught anatomy and physiology at the medical school in Abuzabel, near Cairo in Egypt. He also worked closely with members of the Egyptian royal family and even received the title of 'Bey' (meaning chieftain) in 1839. Thus he had extensive experience observing and interacting directly with extra-European peoples. Hunt included several of Pruner's tables in his work.

• M. Pruner gives the following measures of the bones of the limbs in centimeters.

| Designation of Measures. | Mean Measures. | | | | Individual Measures. | | | | | |
|---|---|---|---|---|---|---|---|---|---|---|
| | Negroes. | | Europeans | | Negroes. | | Europeans. | | | |
| | Males. | Females. | Males | Females. | Man. | Woman. | Man. | Woman. | New-born Infant | Child 5 yrs. old. |
| Total height of Skeleton.... | 160·04 | 148·66 | 172·23 | 164·42 | 160·0 | 156·0 | 160·0 | 157·0 | 42·25 | 101·0 |
| Femur .................. | 44·72 | 42·50 | 47·70 | [illegible] | 45·0 | 41·5 | 45·0 | 42·0 | 6·7 | 23·0 |
| Tibia.................... | 38·09 | 35·33 | 38·76 | 37·71 | 39·0 | 38·5 | 36·0 | 36·0 | 6·0 | 22·0 |
| Length of foot .......... | 24·50 | 21·85 | 25·06 | 23·37 | 23·5 | 21·5 | 24·0 | 23·0 | " | " |
| Humerus................ | 31·27 | 29·50 | 33·72 | 31·57 | 31·5 | 31·0 | 34·0 | 31·0 | 8·2 | 18·0 |
| Radius. ................. | 24·63 | 23·00 | 25·40 | 24·85 | 24·5 | 25·0 | 27·0 | 21·0 | 5·75 | 13·0 |
| Length of hand.......... | 18·64 | 17·00 | 18·84 | 18·14 | 19·0 | 18·0 | 20·0 | 17·0 | " | " |

**Figure 4.4: A table by Pruner showing the differences between the limbs of Africans and Europeans. From J. Hunt, *On the Negro's Place in Nature* (London: Trübner and Co., 1863), p. 8. Reproduced from the author's private collection.**

The purpose of these tables was to provide observable and numerical data for his readers to utilize in their comparative studies. In the case of the table shown in Figure 4.4, the reader was to compare the numerical data on 'Negroes' and 'Europeans' and draw conclusions based on the tabulated material.

Hunt also wanted anthropologists to specialize in specific sub-branches within the sciences of Man, arguing that the resulting research would be more focused and comprehensive. This in turn would help to further develop and refine the research-activities of practitioners. Under this scheme there were four sub-branches. The first was 'archaic anthropology', which looked at the physical remains of past societies and its people. The second was 'historical anthropology', which examined the religions, myths and languages of different ethnic groups. The third was 'descriptive anthropology', which described the anatomy and physiology of the various races that inhabit the world. Finally, the fourth was 'comparative anthropology', which sought to cross-examine different racial groups and demonstrate their commonalities and dissimilarities.[87]

These sub-branches were not autonomous, and specialized practitioners from each group were supposed to work collaboratively in order to produce as extensive a study as possible. Hunt's plan to divide the discipline into more focused sub-branches is interesting because in many respects it looks similar to the modern North American school of anthropology, which divides the discipline into four sub-branches. The first is 'archaeology', which looks at the physical remains of past societies. The second is 'linguistic anthropology', which examines the different languages spoken by ethnic groups around the world. The third is 'socio-cultural anthropology', which investigates and describes different cultures around the world. Finally, the fourth is 'physical/biological anthropology', which looks at the anatomy and physiology of humans.[88]

Another key issue for Hunt had to do with the inclusion of Darwinian evolutionary theory into discussions on humans. Hunt had reservations about any theory that attempted to explain human origins, and challenging Darwin's hypotheses became a key preoccupation for him in the late 1860s. In his 1867 article 'On the Doctrine of Continuity Applied to Anthropology', Hunt wrote that 'All science must be based on facts or on philosophical deductive reasoning'. He continued by stating that he was 'willing to admit that the mutability of species is a fair and philosophical hypothesis', but that it was 'neither warranted by scientific facts, by logical reasoning, nor by philosophical assumption'.[89] For Hunt, Darwinian evolutionary theory – and specifically the mutability of species – was an interesting hypothesis, but it required more observable data to substantiate its claims. The application of Darwinism to the study of human diversity was further problematic because Hunt argued that Huxley and Wallace misrepresented Darwin's work within their own researches.

> There are no facts at present which throw any light on the origin of mankind. Mr. Darwin is only to be blamed because he has not curbed the restless impatience of his disciples. The mischief is being done by Professor Huxley a[n]d Mr. Wallace; but Mr. Darwin has hitherto refrained from coming forward to protest against the abuse which is being made of his theory.[90]

Accordingly, for Hunt, the mutability of species 'may be true when applied to botany or zoology', but there was not enough evidence to support its application in anthropology.[91] Hunt was also critical of ethnological taxonomies because he believed the amount of difference observed among the various races was enough to demark each race as separate species and not varieties.[92]

Hunt was equally critical of 'natural selection', and he argued that Darwin's commitment to slow change over vast periods of time was problematic because it lacked sufficient observable evidence.[93] He asserted that believing in 'natural selection' was no different to believing in 'divine intervention'. He wrote that there was 'no difference between natural selection with unlimited power, and [what] the other calls in a Deity provided in the same manner'. In addition, Hunt stated that

> Mr. Darwin is most successful in his sneers at those who resort to the hypothesis of independent creation to explain the existing species of plants and animals, apparently entirely oblivious to the fact that he lays himself open to exactly the same charge. Men that sneer at special creation must expect in their turn to be treated in a similar manner. Spontaneous generation forms no part of Darwinism. When spontaneous generation is proved to be true, Darwinism will necessarily, by implication, be found to be false ... Future generations will be able to decide on the respective merits of the two systems better than we can do at this time.[94]

Hunt prioritized direct observation, and under this model the Baconian method of induction was strategically used as the cornerstone of his anthropological framework. Observable facts – such as anatomical and physiological data – lay at the foundation of any good study of race. Any theory attempting to explain human origins unsubstantiated by sufficient visible evidence was positioned as erroneous. Hunt's reservations about the application of theories such as 'natural selection' to studies of humans were not distinctive, and Huxley also questioned the value of certain Darwinian principles in ethnological research.

## *Huxley and his Practical Reforms in Ethnology*

During the 1860s, Huxley was steadfast about reforming the intellectual scope and research practices of a broad range of disciplines, including anatomy, physiology, botany, zoology and ethnology. He was particularly concerned about the way in which practitioners utilized 'facts' in their research. As with Hunt, he argued that all scientific studies must be based primarily on directly observable evidence such as anatomical and physiological data.[95] Huxley was also committed to establishing and promoting lab-based research in Britain. He initiated several projects designed to transform the training facilities of researchers and the literature and instruments they used in their studies.[96] Within the ethnological context, one of Huxley's more significant contributions to reforming the discipline's practices was his article 'On the Methods and Results of Ethnology', published in 1865 in the *Fortnightly Review*. It was an instructive guide and aimed to refine the observational practices of ethnologists by building on the works of earlier researchers such as Blumenbach, Prichard and Latham. In the introduction, Huxley defined the research scope of the discipline and wrote,

> Ethnology, as thus defined, is ... the great science which unravels the complexities of human structure; traces out the relations of man to other animals; studies all that is especially human in the mode in which man's complex functions are performed; and searches after the conditions which have determined his presence in the world.[97]

As with Hunt, Huxley also wanted practitioners to conduct more focused studies on human variation. Under this model, researchers specialized in specific sub-branches of the discipline. Huxley divided ethnology into four main areas: 'zoological/anatomical ethnology', which looked at the physical structure of humans; 'linguistic/philological ethnology', which examined the various languages spoken by people around the world; 'material ethnology', which discussed the material culture of both past and present societies; and finally 'historical ethnology', which sought to investigate the history of different ethnic groups, travel reports and historical records.[98] As with Hunt's anthropology, Huxley's sub-branches were to work collaboratively with one another. In doing so, their collective aim was to produce as comprehensive a study of different racial varieties as possible.

Another issue within the sciences of Man that Huxley wanted to deal with was the politics relating to terminology. As we have discussed in the previous chapters, terms such as 'varieties', 'races' and 'species' connected to the professional and political agendas of several key figures. For instance, Prichard and Hodgkin classified and described humans as 'varieties'. In doing so, they were trying to demonstrate that Africans, Europeans and Asians possessed significant overlapping features and thus shared a common ancestral origin. This argument for a single origin linked to Prichard's and Hodgkin's philanthropic and abolitionist agendas. Knox, on the other hand, argued in favour of the multiple origins of humans, and he described different races as 'species'. This polygenetic stance fitted into his vocational strategy.

Huxley, however, wanted to separate all sciences from political and religious discourses and keep the various disciplines ideologically neutral on a range of topics. In the case of ethnology, he wanted to replace the older politically motivated terminology with a new lexicon. He wrote,

> I speak of 'persistent modifications' or 'stocks' rather than of 'varieties', or 'races', or 'species', because each of these last well-known terms implies, on the part of its employer, a preconceived opinion touching one of those problems, the solution of which is the ultimate object of the science; and in regard to which, therefore, ethnologists are especially bound to keep their minds open and their judgements freely balanced.[99]

A secondary reason for Huxley promoting the adoption of these new terms was that it linked ethnology's vocabulary to Darwinism. Part of Huxley's professional strategy was to incorporate Darwinian evolutionary theory into all the natural sciences. However, Hunt immediately recognized that Huxley's argument for an ideologically neutral lexicon was fallacious and that the inclusion of Darwinian terminology into the discipline was part of his career-building strategy. As a response, Hunt identified Huxley's motivating ambitions in his 1866 attack on Darwinism and wrote that 'the term "persistent modification" equally involves a theory on the part of those who use it'. As such, it connected to other larger motivating aims.[100]

There were other similarities between Huxley's and Hunt's observational frameworks. For instance, both figures had reservations about the use of linguistic and philological evidence when substantiating theories about human origin. In Huxley's case, he believed that language was an important component of ethnology, but that researchers using it had to be more critically aware of its limitations. He wanted 'linguistic/philological ethnologists' to be more reflexive about cultural contamination. Was it possible that indigenous peoples' languages may have altered in certain cases due to encounters with other neighbouring populations? If this was the case, the commonalities between the languages were not because the two populations derived from a single origin, but rather because of cultural contact with each other. Huxley wrote,

> it seems to me obvious that, though, in the absence of any evidence to the contrary, unity of languages may afford a certain presumption in favour of the unity of stock of the people speaking those languages, it cannot be held to prove that unity of stock, unless philologers are prepared to demonstrate, that no nation can lose its language and acquire that of a distinct nation, without a change of blood corresponding with the change of language.[101]

Huxley was even prepared to critique Prichard's use of linguistic theory because he believed it lacked a critical analysis of philological evidence. He also claimed that Prichardian ethnology placed more weight on language than it did on anatomy and physiology. Accordingly, Huxley argued that ethnologists should prioritize anatomical and physiological research because it was a more empirically grounded approach to studying humans. He wrote, 'it is plain that the zoological court of appeal is the highest for the ethnologist, and that no evidence can be set against that derived from physical character'.[102] Although Huxley was critical of Prichard's language studies, he did approve of his classificatory practices when organizing humans into groupings based on physical structure. Therefore we can see the influence that figures such as Blumenbach, Prichard and Latham had on Huxley's ethnological taxonomy. Moreover, just as his predecessors, Huxley aimed to further refine the observational practices of ethnologists, and one can see how his training in anatomy and physiology shaped his practical model.

The first step for Huxley was to collect and organize his data. He began by separating the various 'persistent modifications' or 'stocks' inhabiting the world into larger regional groupings, with descriptions of each type. For instance, Huxley described both the physical form and the material culture of Australians, which included Tasmanians, New Zealanders, New Caledonians and Papuans.[103] Because he had travelled to the South Pacific during the *Rattlesnake* voyage, his account was a mixture of his own observations collected *in situ* and those of other researchers and informants. Huxley provided his readers with a general physical description of South Pacific peoples. He wrote, 'Australians have dark, usually chocolate-coloured skins; fine dark wavy hair; dark eyes, overhung by beetle brows; course projecting jaws; broad and dilated, but not especially flattened, noses; and lips which, though prominent, are eminently flexible'.[104] Building upon Prichardian methods, Huxley also discussed the physical conformation of Australian skulls. He stated,

> The Skulls of these people are always long and narrow, with a smaller development of the frontal sinuses than usually corresponds with such largely developed brow ridges. An Australian skull of a round form, or one [where] the transverse diameter of which exceeds eight-tenths of its length has never been seen.[105]

Once Huxley had organized his various 'persistent modifications' or 'stocks' into groupings, the next step for him was to tabulate the material into columns for comparative analysis (see Figure 4.5). Thus, despite Hunt's claim that the Baconian method of induction was a distinctly anthropological practice, Huxley utilized it in his framework. The only difference was that he did not explicitly identify it as Baconian. It was not a central component of Huxley's rhetorical strategy for strengthening ethnology's research claims. Huxley arranged humans into two main categories: 'Leiotrichi', meaning smooth-haired people, and 'Ulotrichi', meaning woolly-haired people.[106]

| | Leiotrichi. | | Ulotrichi. | |
|---|---|---|---|---|
| | Dolichocephali. | Brachycephali. | Dolichocephali. | Brachycephali. |
| Leucous. | | | | |
| | …. Xanthochroi …. | | | |
| Leucomelanous. | | | | |
| | …. Melanochroi …. | | | |
| Xanthomelanous. | *Esquimaux.* | Mongolians. | *Bushmen.* | |
| | | *Amphinesians.* | | |
| | | *Americans.* | | |
| Melanous. | | | | |
| | *Australians.* | | Negroes. *Negritos.* | *Mincopies* (?) |

*** *The names of the stocks known only since the fifteenth century are put into italics. If the "Skrälings" of the Norse discoverers of America were Esquimaux, Europeans became acquainted with the latter six or seven centuries earlier.*

**Figure 4.5: Huxley's 'Persistent Modification' table. From T. H. Huxley, 'On the Methods and Results of Ethnology', *Fortnightly Review*, 1 (1865), pp. 257–77, on p. 269. Reproduced from the author's private collection.**

Under both headings, Huxley further classified his human 'stocks' into groupings based on skull conformation: the 'Dolichocephali' or 'long-headed', and 'Brachycephali' or 'broad-headed'. Within these groupings, he organized his 'stocks' according to their skin, eye and hair colour. The 'Leucous' were people with fair complexions and blond or ginger hair. In cases where a group possessed cranial features from both skull types, Huxley would place them between the two columns such as the 'Xanthochroi' (or 'Nordic people').[107] Huxley's classificatory system based its groupings on those outlined by Prichard in the various editions of his *Researches into the Physical History of Mankind*.[108] However, he further refined the method by visually arranging the data into tables. In doing so, readers could further analyse the material and draw more conclusions. Huxley also provided examples on how to use the tabulated evidence. He wrote, 'It is curious to observe that almost all the woolly-headed people are long-headed; while among the straight-haired nations broad heads preponderate, and only two stocks, the Esquimaux and the Australians, are exclusively long-headed'.[109]

Finally, as with Hunt, Huxley had reservations about Darwin's theory of 'natural selection'. Firstly, he argued that Darwin's hypothesis was excellent, but it required further evidence. Secondly, Huxley remained unconvinced by Darwin's commitment to the principle that '*Natura non facit saltum*' or 'Nature makes no leaps' and argued that the smallest change at the molecular level could create sudden and major transformations in an organism.[110] Therefore Huxley did not staunchly support 'natural selection' as he did other Darwinian theories such as 'transmutationism', but he did encourage its testing. In his 1859 review of the *Origin of Species* (1859), Huxley wrote,

> If it can be proved that the process of natural selection, operating upon any species, can give rise to varieties of species so different from one another that none of our tests will distinguish them from true species, Mr. Darwin's hypothesis of the origin of species will take its place among the established theories of science, be it consequences whatever they may [be]. If, on the other hand, Mr. Darwin has erred, either in fact or in reasoning, his fellow-workers will soon find out the weak points in his doctrines, and their extinction by some nearer approximation to the truth will exemplify his own principle of natural selection.[111]

Although there were clear distinctions between certain aspects of ethnology and anthropology, there were many parallels between Hunt's and Huxley's disciplinary strategies and observational practices. Both utilized a Baconian method of induction for drawing conclusions about human varieties. In addition, both figures argued that the sciences of Man should place more emphasis on anatomical and physiological evidence, and each had reservations about the theory of 'natural selection'.

## Bringing the Data Home: Libraries, Museums and Photography

> It will be the duty of conscientious anatomists carefully to record all deviations from the human standard of organization and analogy with inferior types, which are frequently manifested in the negro race. These observations should be made solely as to the existence of the facts themselves, and without any reference to the theories that may be founded on them. Future generations will thank us more for the establishment of good reliable facts than for any hap-hazard speculations.[112]
>
> James Hunt, 1863

### *Teaching Observational Practices and Hunt's 'Systematic Plan'*

One of Hunt's primary concerns for anthropology was to develop a 'systematic plan' for training informants travelling abroad – and researchers back in Britain – on how to observe humans in a specialized way. In his 1863 'Introductory Address', Hunt stated that 'We have abundance of observations and facts of a certain kind; but the observations are valueless, because nearly all travellers only see what suits their own preconceived notions'.[113] Responding to this problem, Hunt argued that 'For science we must have exact details ... It must be our object to decide what are the facts we most want, and collect information on a systematic plan'.[114] In many respects, Hunt's trepidations about the quality of observations returning to Britain mirrored those of Hodgkin and King, who during the 1840s produced a body of instructive literature that explained how to observe and describe human varieties in a specialized way in order to maximize the value of informants' observations for ethnologists back in Britain.[115]

A secondary concern for Hunt was that many travellers exaggerated their cross-cultural encounters with indigenous people because they wanted to turn their voyages abroad into travelogues that would appeal to a wider readership and not just a specialist audience. Hunt wrote that 'Another cause of the comparative uselessness of the accounts of travellers is the want of honesty in telling what they really saw. Some fear shocking public opinion, while others indulge in exaggerations for the sake of the excitement which their narrative produces in the reading public'.[116] However, Hunt argued that it was not the fault of travellers that their observations collected *in situ* misrepresented or overlooked key areas of concern for researchers back in Britain. Rather, it was the duty of anthropologists to develop an institutionalized system for training people in the sciences of Man. He looked to universities as a key location for anthropological education and wrote, 'it is the duty of our Universities to make the Science of Mankind a special subject of study. I look forward to the day when all our Universities shall have professors whose sole study shall be the philosophy of mankind'. For Hunt, the long-term goal was for the study of human diversity to be 'made part of the national education'.[117]

However, Hunt stated that until there was government aid for founding training courses at universities, it was up to private enterprise to continue developing anthropological practices. He wrote that 'one of the best means of helping to do this is by the establishment of a good and reliable museum'.[118] As we saw in the last chapter, ethnological exhibits were important locations for refining observational practices, and sites such as the ethnological department at the Crystal Palace simulated *in situ* observational experiences. For the ASL, an anthropological museum would provide members with materials to substantiate their research claims. The collections would give researchers access to a large selection of human skulls that fully represented global human diversity. It would also house an assortment of instruments and tools representing the technological prowess of different races. Hunt also wanted to set up an anthropological library so that members could read instructional literature on how to observe and analyse human variation and look at travel accounts.

Hunt and the ASL council routinely updated members about the progress of the society's museum and library. In 1864 the council reported that

> At present the [museum] specimens, many of which are of the highest possible scientific value, and one or two unique in European collections, have been placed temporarily in a small room adjacent to the Library. The Council hope that the increased finances of the Society during the year 1865 will warrant such expenditure upon the necessary fittings and furniture of the Museum as will be requisite to exhibit adequately the whole of the collection.[119]

The council reported that the library possessed 'more than 500 volumes, which have been all bound and stamped', and that 'A catalogue is in course of preparation, which will be arranged both according to subjects and alphabetically'.[120] By 1867 Hunt further reported that 'In the last five years we have collected together one hundred and five skulls, beside a large number of flint, stone, and other weapons and implements of ancient and modern races of man. This forms the nucleus of further collections'.[121]

However, it was not enough to simply provide researchers with collections to work with; it was also important to train them on how to use the material and maximize their results. Thus Hunt organized a lecture series at the ASL that aimed to improve the observational practices of anthropologists. In his 1867 address, Hunt outlined the significance of these disquisitions, stating that

> These lectures may also sometimes take the form of oral instruction. How many of our Fellows, for instance, would not be glad to have instruction in the employment of the different craniometers now in use, or even on craniology generally? How many more would not gladly listen to a practical descriptive anthropologist, like Dr. John Beddoe, while he discoursed on his system of making observations? Or who would not be glad to attend a course of lectures by such men as Captain Burton ... on races with whom [he has] come in contact?[122]

Hunt also organized a scheme to establish a network of informants (whom he called 'local secretaries'), whose job was to collect ethnographic data in the colonies and send it to the ASL for researchers to use in their studies. Hunt stated that 'there are many who are anxious to render some assistance to science, but do not know what to do, as they are ignorant [about] what sort of information science requires. If our local secretaries are carefully selected, and proper questions and instructions are sent to them, I look forward with much hope to the benefit that will accrue to science from such a plan'.[123]

Hunt was also involved in a vigorous publishing scheme that translated continental works into English so that British ethnologists and anthropologists could incorporate these continental methodologies into their studies. This included books such as Paul Broca's *On the Phenomena of Hybridity in the Genus Homo* (1864), George Pouchet's *On the Plurality of the Human Race* (1864) and Carl Vogt's *Lectures on Man* (1864).[124] Thus Hunt aimed to improve the observational practices of researchers and informants by establishing different kinds of resources for them to utilize in their studies. This included the formation of an anthropological museum and library, the creation of a specialized lecture series on human variation topics, and a scheme to develop a large informant network. Taken together, these initiatives demonstrate another instance where nineteenth-century anthropology aimed to transform its research practices and bring its 'object of study' from abroad to the metropole.

## *Huxley's 'Well-Considered-Plan' and Anthropometric Photography*

Hunt was not alone in organizing schemes to bring ethnographic materials to Britain for researchers to utilize in their studies. Huxley aimed to systematize the collecting practices of informants living abroad by formulating a set of instructions on how to photograph indigenous people for ethnological research. The plan emerged in 1866 when Huxley received a letter from Sir Joseph Fayrer (1824–1907), who was a physician living in Calcutta. Fayrer invited Huxley to visit the region and participate in a project that aimed to photograph the different native tribes of the area. Unfortunately, Huxley declined the offer because of his lecturing commitments in Britain. However, he was very interested in the project, calling it 'a grand and most useful one', and he provided Fayrer with some suggestions on how to make the plan work:

> It strikes me that it will not do to be content with a single representation of each tribe. At least four or five will be needed to eliminate the chances of accident, and even then much will depend upon the discretion and judgement of the local agent who makes the suggestion. This difficulty, however, applies chiefly if not solely to physical ethnology. To the philologer the opportunities for comparing dialects and checking pronunciation will be splendid, however [few] the individual speakers of each dialect assembled [are] ...[125]

Furthermore, Huxley told Fayrer, 'You will want, I should think, a physical and a philological committee to organize the schemes: (1) for systematic measuring, weighing, and portraiture, with observation and recording of all physical characters; and (2) for uniform registering of sounds by Roman letters and collection of vocabularies and grammatical forms upon an uniform system'.[126] Huxley hoped that Fayrer's project would furnish 'linguistic/philological ethnologists' and 'zoological/anatomical ethnologists' in Britain with data for their researches. In the same letter, Huxley also warned Fayrer not to contact Hunt and the ASL about the project, stating, 'don't have anything to do with the quacks who are at the head of the "Anthropological Society" over here. If they catch scent of what you are about they will certainly want to hook on to you'.[127] We can clearly see Huxley's professional rivalry with Hunt coming to the fore.

Huxley's involvement in photographing ethnological specimens around the world increased as the 1860s came to a close. After he was elected president of the ESL in 1869, one of his first schemes was to formulate a set of instructions for colonial informants on how to objectively capture human 'stocks' for scientific photographs. Huxley lamented in a private note that

> Great numbers of ethnological photographs already exist but they lose much of their value from not being taken upon a uniform and well-considered-plan. The result is that they are rarely either measurable or comparable with one another and that they fail to give that precise information respecting the proportions and the confirmation of the body alone are of any considerable worth to the ethnologist.[128]

In order to facilitate the success of his photographic project, Huxley required the support of the British government. Thus in 1869 he sent a letter to the Colonial Office in London, requesting their support in getting their officers to assist him in systematically photographing the indigenous populations of the colonies. Huxley positioned the project as being beneficial for the empire because the images would help ethnologists back in Britain produce what he believed to be more objectively accurate reports on the people who inhabited the colonies. In turn, this would provide colonial administrators with information that would help them better manage their imperial settlements.[129] On 30 November 1869 the Colonial Office of London sent a notice to all its colonial governors asking them to photograph the races of the empire for the advancement of science. The request contained further instructions explaining exactly how to arrange the models for photographing.[130]

According to Huxley's set of instructions, colonial informants were to photograph their models standing unclothed, facing forward and in profile, with their hands not obstructing their bodies in any manner. Their ankles were to be touching (see Figures 4.6 and 4.7).[131] He requested that an adult male, an adult female and a child of either sex be captured from each human 'stock' photographed. He also provided colonial administrators with precise specifications for the distance at which the figures were to stand from the camera and asked that a measuring stick be placed in the photograph to provide researchers with a sense of scale.[132]

Overall, forty sets of images (or one hundred and sixty photographs) following Huxley's instructions returned to Britain. In many cases, colonial officials struggled to find models willing to be photographed naked.[133] For instance, the governor of British Columbia, Anthony Musgrave (1828–88), reported that

> I am informed that no Indians here will consent to be photographed in a state of nudity, although reward has been offered. It is believed that they have a superstitious dread of some hidden purpose which they do not understand and it would [be] impossible to explain to them the scientific object of the proceeding.[134]

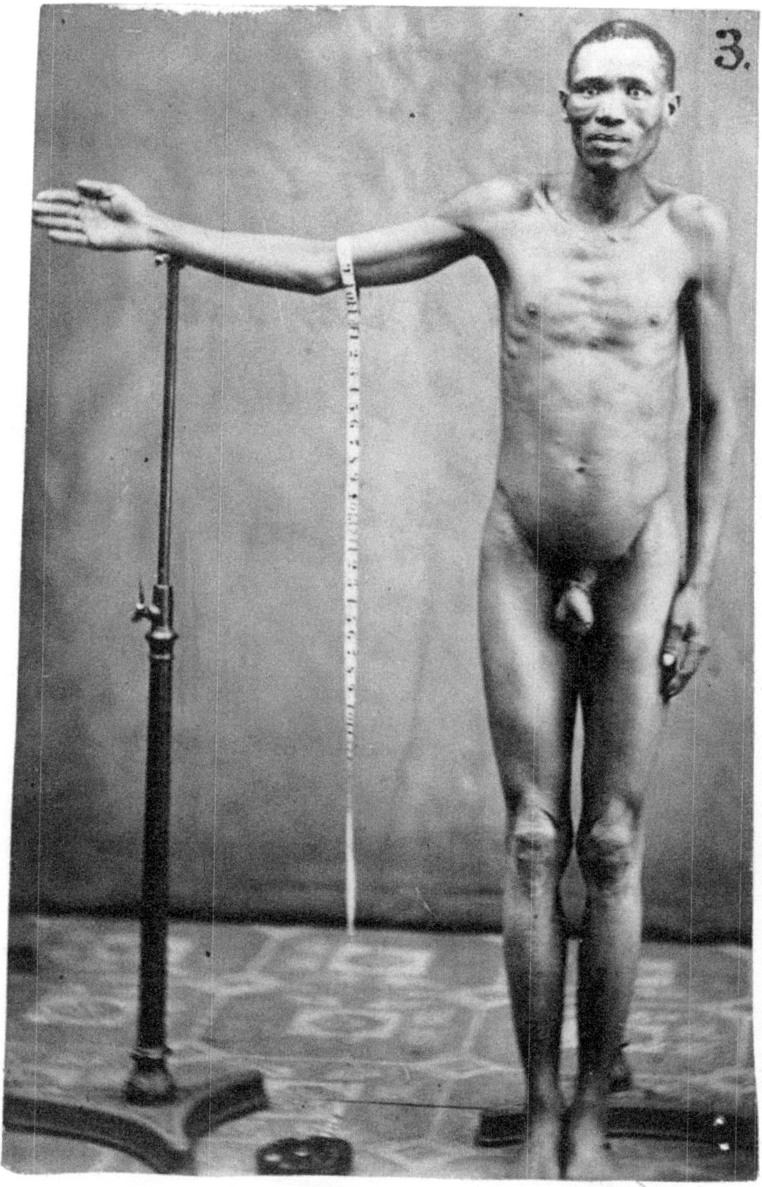


Figure 4.6: A full-length and full-face anthropometric photograph of Abraham Lucas, a South African 'Hottentot', taken according to Huxley's instructions. Reproduced by permission of the Pitt Rivers Museum, University of Oxford, catalogue no. PRM 1998.211.13.2.

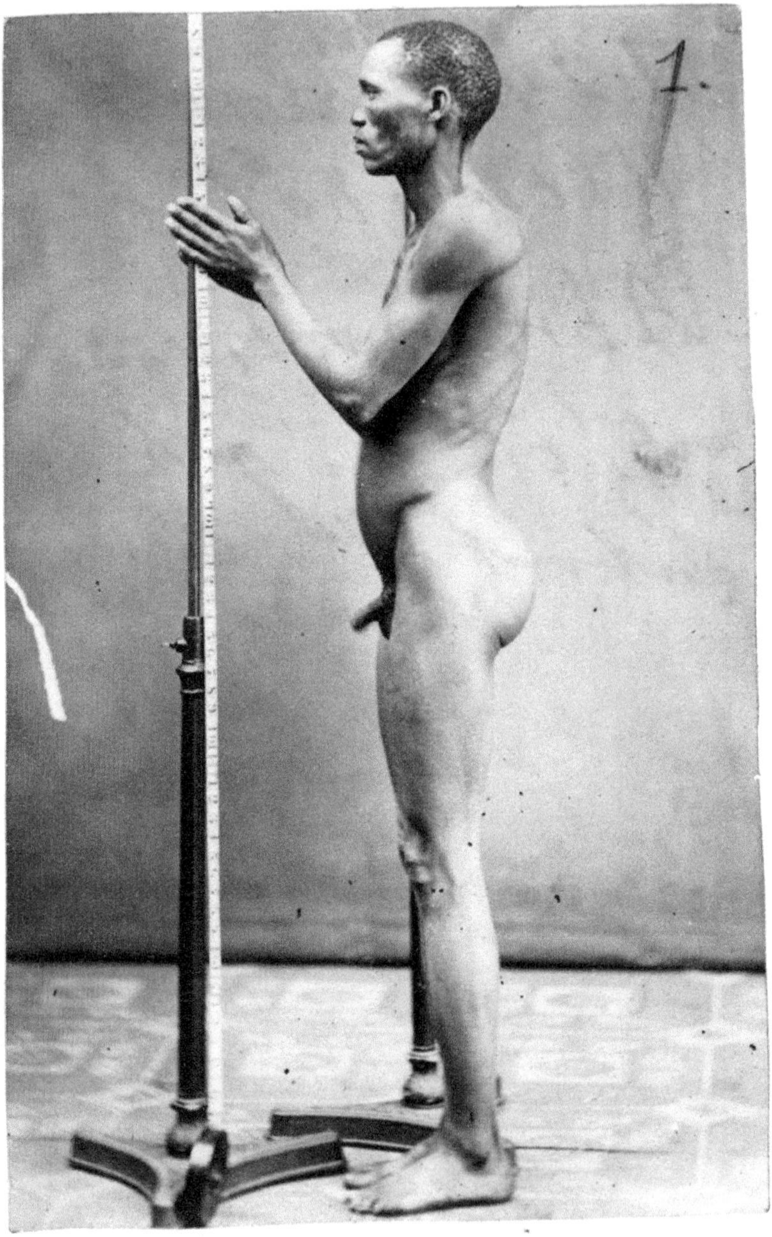


Figure 4.7: A full-length profile anthropometric photograph of Abraham Lucas, a South African 'Hottentot', taken according to Huxley's instructions. Reproduced by permission of the Pitt Rivers Museum, University of Oxford, catalogue no. PRM 1998.211.3.

Huxley's scheme was a highly exploitative practice, and many of the photographs that were collected were of individuals who had lost any right to protest against this activity. Several of the images that returned from South Africa were of convicts stationed at the Breakwater Prison in Cape Town.[135] However, Huxley was not the only member of the British ethnological community interested in photographing human varieties, and in the same year that he became president of the ESL, J. H. Lamprey published a report on how to photograph humans for ethnological research.[136]

Huxley was also promoting the importance of building colonial networks at the ESL, and he encouraged travellers to assist the sciences of Man by collecting ethnographic data. He wrote in his 1869 presidential address that 'we earnestly invite the co-operation of persons who have been, or are likely to be, residents in the countries [throughout the empire]'. Huxley continued by stating that

> We hope that the co-operation we seek will take two forms. On the one hand, we trust that those who, as old residents, possess information, will give it to us for the benefit of the public. And, on the other hand, that those who are going to be residents abroad will attend for the purpose of learning how easy for them it is to serve science and forward the solution of great and interesting problems by the expenditure of a small amount of thoughtful and intelligent attention.[137]

Huxley hoped that travellers would begin to carefully observe the indigenous people they came into contact with and send 'thoughtful' and 'intelligent' reports back to Britain for researchers to use. It was another instance where his practical model for transforming the discipline overlapped with Hunt's programme.

## Conclusion: Towards the 'Anthropological Synthesis'

On 29 August 1869 James Hunt died at his home in Hastings, East Sussex, at the age of thirty-six. It was a quiet end to his vocal career as a prolific disseminator of science. In 1870 the *AR* republished an obituary from the *New York Weekly Day Book* (from 6 November 1869) stating, 'We are pained to hear of the death of Dr. Hunt, the founder of the London Anthropological Society and editor of the London *Anthropological Review*, beyond doubt, or, at all events, the most useful man in England'.[138] It was a gleaming obituary for one of the most controversial scientific figures in Britain during the 1860s. After all, in the closing years of the decade, opposition to Hunt and his activities grew, and it took a toll on his health. He had gained a reputation within the scientific community as an argumentative and ill-tempered individual. Evidently, the *AR* wanted to ensure a positive and lasting reputation for its editor as a central propagator and reformer of the sciences of Man in Britain, rather than as a troublemaker.

A report in the *British Medical Journal* by E. S. Collingwood positioned Hunt 'amongst the martyrs of science'.[139] According to Collingwood, Hunt died from over-exhaustion while fighting for an anthropological subsection at the BAAS meeting in Exeter. Collingwood wrote, 'that a man should engage in the battle for what he believes truth with more vigour than his bodily strength will permit is rather a credit than a disgrace'.[140] Thus Collingwood argued that Hunt's health was second to his commitment to anthropology, demonstrating his dedication to reforming the science. His death was a shock to his friends and colleagues at the ASL. The newly elected president John Beddoe (1826–1911) – in a special address – described Hunt's prolific career as a respected physician and founder of British anthropology. Beddoe stated, 'In the study of the vocation he had chosen, he displayed the zeal and energy which characterized him in all his undertakings: he collected a complete library of works, English and foreign, bearing on the various branches of the subject; made numerous and valuable independent observations and improvements ... and embodied the results of these studies in several published works'. Beddoe recognized the important role Hunt played in establishing an anthropological library and improving the discipline's theories and practices. With Hunt gone, Huxley was eventually able to seize control of the discipline and firmly place Darwinism at its centre.[141]

My purpose in this chapter was to reconsider Hunt's and Huxley's contributions to the scientific race studies during the 1860s. Historians and anthropologists who have written about this decade have tended to position the rivalry between our two figures as a competition over disciplinary practices. However, there were significant similarities between Hunt's and Huxley's practical models. The 'anthropological controversy' of the 1860s was a professional competition for the leadership of the research field. Looking at Hunt's and Huxley's educational backgrounds and their vocational strategies for obtaining the leadership of the discipline is illuminating. It demonstrates how both figures saw training in anatomy and physiology as essential for scientific observation, and how each of them was promoting a form of 'scientific naturalism' during a period when older explanatory models of the world were being replaced by new ones. Section two re-examined the causes that have customarily been associated with Hunt's exit from the ESL, and linked it to his observational framework and the founding of the ASL. In addition, it demonstrated that Huxley had similar views to Hunt when it came to both the woman and race questions. Finally, sections three and four compared the observational practices of Hunt and Huxley and demonstrated the commonalities between the two figures' observational models. Both used a Baconian method of induction when examining ethnographic data. Hunt and Huxley also organized colonial networks of exchange that aimed to bring a higher quality of ethnographic material back to Britain for researchers to incorporate into their studies on humans.

As the decade of the 1860s came to a close, more changes in disciplinary practices emerged. With Hunt's death, Huxley was able to broker a deal between the ESL and the ASL to amalgamate the two societies. In 1871 he helped found the Anthropological Institute of Great Britain and Ireland. However, to satisfy members of the ASL, ethnologists agreed to call the discipline anthropology, and in return Lubbock (who was a leading member at the ESL) was made the first president of the newly formed institute. In the same year, Darwin published his *Descent of Man* and Tylor published *Primitive Culture*. As we shall see in the following chapter, both texts were important for the study of humans because each book aimed to further refine the observational practices of anthropologists and firmly place the discipline on an evolutionary model.

# 5 SYNTHESIZING THE DISCIPLINE: CHARLES DARWIN, EDWARD BURNETT TYLOR AND DEVELOPMENTAL ANTHROPOLOGY IN THE EARLY 1870s

> During many years I collected notes on the origin or descent of man, without any intention of publishing on the subject, but rather with the determination not to publish, as I thought that I should thus only add to the prejudices against my views.[1]
>
> Charles Darwin, 1871

The year 1871 was an important one for the discipline of anthropology. It marked the beginning of a new era for the young science. After eight years of warring between ethnologists led by Thomas Huxley and anthropologists led by James Hunt, the two sides had finally resolved many of their differences and agreed to amalgamate and form the Anthropological Institute of Great Britain and Ireland. Both sides were forced to make compromises. The ethnologists agreed in principle to accept the designation of 'anthropologist' as their new name, so long as the president of the Anthropological Institute was John Lubbock. The hope was that under the leadership of one of the leading scientific naturalists of the mid-Victorian period, the sciences of Man would continue to move forward on sound empirical principles.[2]

In the same year, however, there were also two important books published on anthropological topics which taken together helped to firmly place the discipline onto a more rigorous developmental framework. One of these books was Darwin's *Descent of Man, and Selection in Relation to Sex,* which he claimed was his first major contribution to discussions on human variation. As the quote above suggests, throughout the anthropological schism of the 1860s, Darwin distanced himself from the harsh criticisms he received from figures such as Hunt. The American Civil War (1861–5) further augmented these critiques because questions relating to the unity of races and human origin came to the fore of public debates. However, with Hunt's untimely death in 1869 at the age of thirty-six, and the conclusion of the American conflict, Darwin finally broke his silence and publicly entered into discussions on human evolution in the early 1870s.[3]

The other significant text published in 1871 was Edward Burnett Tylor's *Primitive Culture: Researches into the Development of Mythology, Philosophy, Religion, Language, Art and Custom*. It was Tylor's second major contribution to the sciences of Man, and it demonstrated the maturation of many of the theories and methods discussed in his earlier work. Tylor's aim was to formulate a system that organized non-physical pronouncements of different societies – such as their religions, customs and languages – into a developmental scheme. It was a model built upon older Prichardian ethnological techniques that attempted to unite the different tribes and nations of the world to a common origin through philological comparisons.

This chapter seeks to enlarge our understanding of anthropological observational practices at the beginning of the 1870s. In certain instances, aspects of this analysis will be familiar. However, within the context of this book, my aim is to connect this well-known story to my discussion on observational practices. In doing so, I hope to add a new dimension to an older historiographic narrative. Traditionally, scholars have described this period as the 'evolutionist era', framing it as a decisive moment in the history of the discipline. There has been a predilection to separate the 1870s from the earlier ethnological period and position it as the starting point of modern anthropology. George Stocking, for example, wrote that 'For several generations of anthropologists, and many still today, the work of E. B. Tylor marked a disjunctive founding moment in the history of anglophone [*sic*] anthropological thought'.[4] Accordingly, attempts by practitioners such as Darwin and Tylor to place the discipline on a sound evolutionary framework have been contextualized in the secondary literature as a novel methodological approach to studying humans.

However, in making the point that evolutionary processes became a new preoccupation for anthropologists in the 1870s, the historiography has underemphasized the significance of developmental processes for Prichardian ethnology.[5] Therefore this chapter's primary aim is to examine the evolutionary models of Darwin and Tylor in turn, and to demonstrate how their work built upon and transformed the pre-existing practices and activities of ethnologists and anthropologists from the first half of the nineteenth century. For many anthropologists in this period, evolution was positioned as a more refined and secular version of monogenism. This chapter will position the 1870s as a period of disciplinary synthesis, bringing together techniques from many of the works already discussed in the previous chapters.

There are two important commonalities between Darwin and Tylor. Firstly, in each case, both figures discussed Man's slow progression from 'savage states' to 'civilized societies' and attempted to show the various stages humans undergo as they develop from simple to more complex organisms. Secondly, both Darwin and Tylor had first-hand experience observing different human varieties *in situ* while travelling abroad, and this provided them with important data to incorporate into their later anthropological work. In section one I will look at the early careers of Darwin and Tylor and examine the impact their travel experiences had on their observational practices. How researchers made sense of their ethnographic data was an important part of their observational practices. Thus in section two I will discuss in detail the analytical frameworks outlined in Darwin's *Descent of Man* and Tylor's *Primitive Culture*, showing how both figures built upon and transformed the theoretical and methodological models of earlier researchers.

Finally, in section three I will shift my focus to questionnaires. Firstly, I will examine Darwin's instructions on emotions, which he sent to informants in 1867. I will show how the responses he received from these queries formed the basis of his book *The Expression of Emotions in Man and Animals* (1872). Secondly, I will look at a small manual from 1872 entitled 'Questions for Explorers (With Special Reference to Arctic Exploration)', which was designed by several members of the Anthropological Institute, including Tylor. Stocking noted that 'Tylor was throughout his career concerned with issues of method, and one assumes that a decade of further reflection in the context of his correspondence with such observers on-the-spot … contributed to [his] heightened ethnographic sophistication'.[6] Although 'Questions for Explorers' is a fairly obscure text within the historiography today, at the beginning of the 1870s it was an important precursor to *Notes and Queries on Anthropology*, which was the most famous ethnographic questionnaire ever to be produced by anthropologists, and was reworked into six different editions from 1874 right up until 1951. Examining these questionnaires will help to further illustrate how both of our figures were building upon earlier disciplinary practices. Rather than positioning the mid-nineteenth century as a period of innovation within the research field, it is more instructive to characterize the late 1860s and early 1870s as a period of synthesis within British anthropology.

## Observing Man *in Situ*

> In the spring of 1856, I met Mr. Christy accidentally in an omnibus at Havana. He had been in Cuba for some months, leading an adventurous life … and visiting all sorts of people from whom information was to be had …[7]
>
> Edward Burnett Tylor, 1861

### *Charles Darwin and his Encounter with the 'Savage'*

Charles Darwin is one of the most studied British men of science from the nineteenth century, and although there is a tremendous amount of historiography about his career, it is striking that very little has focused on his contributions to anthropological theory and practice. When Darwin is discussed in the secondary literature, he is usually positioned as an outsider or a naturalist who influenced the discipline by providing ethnologists and anthropologists with a more rigorous theoretical foundation for their monogenetic arguments, which did not capitulate to biblical explanations.[8] This characterization is suggestive when we examine Darwin's anthropological activities in the 1870s because *Descent of Man* was essentially a natural history work that considered questions of relevance to the sciences of Man. However, in the 1830s Darwin was an important member of the British ethnological community, actively involved in contributing valuable ethnographic observations from his travels and refining the discipline's methods. Moreover, Darwin's comprehensive practical training in medicine and natural history provided him with a sound foundation for writing about human variation.

During the 1820s Darwin attended the medical school in Edinburgh, where he studied human anatomy, physiology and natural history. His time in Scotland had a significant impact on his later ethnological and anthropological work. The Scottish capital was an important place for ethnological training because it provided students with both a sound theoretical background and hands-on experience working with patients and cadavers.[9] Several leading ethnologists, including Prichard and Thomas Hodgkin, acquired their medical education at this historic university. The city also boasted several important museums and extramural schools, such as John Barclay's anatomical school, which had some of the most eminent anatomical instructors of the day, including Robert Knox.[10] The 1820s was also a period when craniology was at the forefront of medical discussions, and figures such as George Combe (1788–1858) regularly demonstrated to audiences a method of conducting phrenological observations.[11] Even though Darwin was not well suited for medicine – because he could not stomach the dissection room – it was still an important developmental period for him.

Darwin took advantage of some of the many scholarly resources available in Edinburgh and, for instance, enrolled in Robert Jameson's natural history class. Jameson (1774–1854) was a renowned geologist and zoologist who had conducted field studies in many of the northern Scottish islands, such as the Shetlands, as well as parts of Ireland. He was also the curator of the Natural History Museum in Edinburgh and played a key role in expanding and improving its collections. By the time of his death in 1854, the museum possessed over 40,000 geological specimens, 10,000 fossils and 8,000 zoological and ethnological objects, including a range of human skulls.[12] Using these skulls, Jameson's natural history lectures included discussions on human variation. This was significant because Darwin would have learned discriminating observational practices that taught him how to identify important characteristics on different types of human crania. This was a valuable methodology to incorporate into ethnological research.[13]

Adrian Desmond and James Moore argue that classes such as Jameson's were designed to train young medics and naturalists on how to conduct natural history research when travelling, and they wrote that 'Even if Darwin wasn't actively preparing for travel, he was prepared for it'.[14] Among the many books on Darwin's lengthy reading list from 1826 were several travelogues, which provided early ethnologists with important ethnographic data and examples on how to describe encounters with indigenous populations.[15] Darwin further enriched his education by studying natural history at Cambridge under the tutelage of the botanist and geologist John Stevens Henslow (1796–1861).[16] In 1830, when Captain Robert Fitzroy was looking for a 'well-educated and scientific person' to be his naturalist on board the HMS *Beagle*, it was Henslow who recommended Darwin for the post.[17]

On 27 December 1831 Darwin set sail on board the *Beagle*. Throughout his five-year journey, he had many opportunities to examine different human varieties in their indigenous habitats, and he recorded his experiences in his travel diary. One can see how Darwin's training in medicine and natural history fed into his observations on humans, and this would later influence his anthropological writings. His journal also provided other ethnologists back in Britain with invaluable ethnographic data to incorporate into their research. It was a significant resource for these researchers because it documented the experiences of someone who had seen extra-Europeans first-hand. However, Darwin's diary entries were not ideologically neutral, and his views about race and the treatment of slaves in the colonies were clearly visible in its pages. It was not an objective representation of his travels. The accounts recorded were full of personal insights. As with many researchers interested in the sciences of Man during the 1830s, Darwin believed in the unity of races.[18] After all, he was raised by a family who staunchly supported the abolition of slavery. Both of his grandfathers were leading supporters of William Wilberforce's emancipation campaigns from the late eighteenth century, and as he was growing up, Darwin's sisters introduced him to several treatises arguing for racial equality. Thus the seeds of his monogenism were sewn when he was young.[19]

Darwin's ethnological interests were a frequent theme throughout his journals. In one of his earliest encounters with extra-Europeans on the island of St Jago off the coast of Africa, Darwin recognized the intelligence of the Africans with whom he came into contact. He wrote, 'I never saw anything more intelligent than the Negros, especially the Negro or Mulatto children. – they all *immediately* perceived & are astonished at the percussion drums. – they examine every thing with the liveliness attention, & if you let them the children chattering away, will pull everything out of your pockets to examine it'.[20] This was not an isolated remark, and Darwin regularly noted similarities between the various races he observed in his travels. Interestingly, there were recognizable examples of Prichardian monogenetic theorizing in his journal entries. For example, while studying the local tribes on the island of Chiloe off the western coast of South America, he wrote, 'Everything I have seen convinces me of the close connection of the different tribes, who yet speak distinct languages ... It is a pleasant thing in any case to see aboriginal inhabitants, advanced to the same degree of civilization, however low that may be'.[21]

There are many other examples of Darwin comparing different racial groups and linking them to a common origin. During his travels in the South Pacific, for instance, he wrote, 'Looking at [the] New Zealander, one naturally compares him with the Tahitian; both belonging to the same family of mankind'.[22] He continued by providing a descriptive analysis contrasting their physical appearances, and he stated that 'Their figures are tall and bulky, but in elegance are not comparable with those ... in Tahiti'.[23] Darwin also attempted to place the different tribes he observed *in situ* onto a developmental spectrum, and this built upon earlier methodologies. Lawrence had discussed the development of races according to successive stages in his lectures on the *Natural History of Man* from 1819.[24] The difference, however, was that Darwin's ideas were based on his first-hand experience seeing the different populations *in situ*. He wrote, 'If the state in which the Fuegians [*sic*] live should be fixed on as zero in the scale of government, I am afraid New Zealand would rank but a few degrees higher, while Tahiti even as when first discovered, would occupy a respectable position'.[25]

Although Darwin attempted to formulate a rudimentary system for ranking indigenous people according to the state of their social organization, he was taken aback by the vast difference between European society and the 'savage' state. In his famous encounter with the indigenous people of the Tierra del Fuego, he wrote,

> When we landed [at the Tierra del Fuego] the party [of four Fuegians] looked rather alarmed, but continued talking & making gestures with great rapidity. – It was without exception the most curious & interesting spectacle ever beheld. – I would not have believed how entire the difference between savage & civilized man is.[26]

Darwin's five-year journey around the world influenced his writings on human diversity. It provided him with an opportunity to observe *in situ* several different types of human varieties.[27] Throughout his many encounters with indigenous peoples, he regularly recorded their physical appearance and made rudimentary monogenetic observations.

Upon his return to Britain in 1836, Darwin continued to think about ethnological questions, and his notebooks show that he was reading the works of leading practitioners interested in the sciences of Man, such as Prichard, Lawrence and Johann Friedrich Blumenbach. In notebook C, written between February and July of 1838, Darwin stated, 'The *races* of men differ chiefly in

~~size~~ colour, form of head & features (hence intellect? & what kinds of intellect) quantity & kind of hair forms of legs – hence the father of mankind probably possessed a structure in these points for a less time than other points ... make [an] abstract on this subject from Lawrence, Blumenbach & Prichard'.[28] Darwin wanted to strengthen his comparative analysis of human variation by consulting the works of top ethnological researchers.

Darwin also believed that many of the populations he encountered during the *Beagle* voyage were destined to die out because of disease and exhausted natural resources from colonial settlements encroaching on indigenous peoples' lands. As was the case with many ethnologists in the 1830s, Darwin supported a plan to systematically record as much data as possible on these races before it was too late. At the BAAS meeting in 1839, Darwin – along with other leading ethnologists, including Prichard – put together an ethnographic manual on how to observe and collect data on human varieties, entitled *Queries Respecting the Human Race to be Addressed to Travellers and Others*.[29] Its aim was to assist informants in documenting the races of the world that were most susceptible to extinction before they disappeared.

Early in his scientific career, Darwin engaged in ethnological research and was actively involved in collecting ethnographic data and developing and refining the research field's observational practices. However, for several decades after the publication of the ethnographic manual in 1842, Darwin abstained from publicly discussing ethnological topics. This was partially due to his recurring health problems, but also because he was focusing on his natural history writings on plants and animals.[30] Moreover, unlike figures such as Huxley and Hunt, Darwin was not vying for the leadership of the discipline. Because of his family's wealth, he was able to self-finance his research without having to take up paying posts in London or elsewhere, or compete for honorific positions within learned societies.

## *Edward Burnett Tylor and his Journey through Mexico*

Anthropologists have traditionally canonized Edward Burnett Tylor as the father of modern anthropology, referring to the discipline in the nineteenth century as 'Tylor's science'. Some have claimed that his contributions form the basis of many of the cultural theories still used by researchers today.[31] Tylor has been positioned as the great teacher of 'civilization', a term that was interchangeable with 'culture'. He has been credited as the first practitioner to define the word, a pioneer in sending his students into the field, and the forerunner to devote his career strictly to the study of human variation. His impact was already being recognized throughout the second half of the nineteenth century and into the early twentieth century. In 1907 his close friend and fellow anthropologist Andrew Lang (1844–1912) argued that 'he who would vary from Mr. Tylor's ideas must do so in fear and trembling (as the present writer knows from experience)'.[32] While

Tylor certainly played a key role in developing the methods and theories of the discipline, his work was not the first of its kind. Rather, he was an astute synthesizer who brought together and refined the techniques and analytical practices of researchers from both sides of the anthropological debates of the 1860s, as well as Prichardian methods from the earlier decades. Furthermore, Tylor's rise to prominence coincided with a period of major reform within the sciences of Man, when the newly amalgamated community was searching for fresh leadership.

Before Tylor obtained his apparent hegemonic position within the discipline, he followed a slightly different training regime from the researchers discussed in previous chapters. Unlike ethnologists such as Latham and Lawrence, or anthropologists such as Hunt and Charles Carter Blake, Tylor did not formally study medicine or natural history at university. He was born into a middling-sort Quaker family and received most of his education from the Grove House School in Tottenham, which was operated by the Society of Friends. Many prominent ethnologists in the first half of the nineteenth century, including Prichard and Hodgkin, were Quakers, and their religious beliefs influenced several of the discipline's theories. The Quaker doctrine of the 'inner light' in particular placed great emphasis on viewing all humans as equal, and this aligned easily with monogenism.[33] From the opening pages of Prichard's *Researches* from 1813, there were discernible religious undertones, and this was a motivating influence throughout the various editions of his work.[34] Thus early on Tylor was immersed in a culture where issues relating to the potential for racial parity were at the forefront of communal discussions.

However, it was Tylor's travel experiences in the 1850s that principally shaped the foundation of his ethnological and anthropological writings. His initial interest in ethnology and natural history came from his older brother, Alfred Tylor (1824–84), who was an archaeologist, geologist and brass founder. Alfred had visited the United States and took up an interest in archaeology and geology. In 1846 he was elected a fellow of the Geological Society and was close friends with some influential naturalists such as Edward Forbes.[35] According to Stocking, it was Alfred who introduced Tylor to ethnological topics and persuaded him to go abroad. After the death of his parents in 1852, Tylor began working in his family's foundry and within a few years developed tuberculosis, which forced him to change careers. Alfred encouraged his younger brother to visit North America to clear his lungs. Taking his brother's advice, in 1856 Tylor set out on a two-year trip.[36]

In the introduction of his travelogue, entitled *Anahuac: or Mexico and the Mexicans, Ancient and Modern* (1861), Tylor wrote that he spent 'the best part of a year' travelling down the Mississippi River and observing the Native Americans and the African slaves he encountered along the way. He also lived for a short time on a sugar plantation in Louisiana, before deciding to visit Cuba for a new adventure.[37] It was a chance encounter with the ethnologist Henry Christy in Havana, however, which really brought Tylor's interests in human variation to the fore. There was much in common between Christy and Tylor. Both were Quakers, grew up in London, and came from middling-sort families. Moreover, they had similar educational backgrounds, having studied at schools operated by the Society of Friends. With so much in common, they quickly established a strong rapport. Christy would become Tylor's mentor over the course of the next ten years until his death in 1865. He would teach Tylor how to observe ethnological specimens *in situ*, as well as explain the major tenets of Prichardian monogenism. When the two first met in 1856, Christy invited Tylor to accompany him on a four-month horseback journey through Mexico.[38]

The aim of Christy and Tylor's excursion through Mexico was to collect as much ethnographic information as possible on the indigenous peoples. Once collected, this material would then be organized, analysed and reworked into ethnological studies on Mexicans. In the opening pages of his travelogue *Anahuac*, Tylor wrote,

> The journey and excursions in Mexico which have originated the narrative and remarks contained in this volume were made in the months of March, April, May, and June of 1856, for the most part on horseback. The author and his fellow-traveller enjoyed many advantageous opportunities of studying the country, the people, and the antiquities of Mexico, owing to the friendly assistance and hospitality which they received there. With this aid they were enabled to accomplish much more than usually falls to the lot of travellers in so limited a period; and they had the great advantage too, of being able to substantiate or correct their own observations by the local knowledge and experience of their friends and entertainers.[39]

Tylor emphasized how he and Christy immersed themselves in the Mexican culture, basing their analytical understanding of the society on 'local knowledge and experience'. In many ways this can be interpreted as a form of participant observation *avant la lettre* because they were trying to see the world through the eyes of the local population. Moreover, we can begin to see how Tylor developed his observational programme by engaging directly with his object of study *in situ*.

From the onset in his travelogue, Tylor noted that Mexico was a remarkable place to conduct ethnological research. He argued that the substantial amount of archaeological evidence available throughout the countryside made it possible to trace the history of the indigenous peoples. Because many of these ethnographic materials were left untouched for several centuries, they were valuable resources for ethnological museums back in Britain. When visiting an Aztec site on one occasion, Tylor wrote that 'Everywhere the ground was full of unglazed pottery and obsidian; and we even found arrows and clay figures good enough for a museum'.[40] Tylor recognized that his primary experience travelling through Mexico and seeing the landscape through his own eyes reshaped his understanding of the region's history. He stated that many British ethnologists were wrong in disputing the credibility of the Spanish reports from the sixteenth century. In fact, Tylor believed that the Spanish chroniclers, if anything, underemphasized the affluence of the Aztec society. He wrote,

> When we left England, we both doubted the accounts of historians of the Conquest, believing that they had exaggerated the numbers of the population, and the size of the cities, from a natural desire to make the most of their victories ... But our examination of Mexican remains soon induced us to withdraw this accusation, and even made us include to blame the chroniclers for having had no eyes for the wonderful things that surrounded them.[41]

Tylor did not just look at materials from his saddle as he passed through an archaeological site; he would dismount from his horse and engage directly with the objects. When he visited the pyramids near Micaotli, he wrote that he 'sat cross-legged on the ground' while members of the local community 'brought many curious articles in clay and obsidian' to examine.[42]

In addition to writing about the history of the Mexican people, Tylor also discussed at length their contemporary state. He was surprisingly reflexive in his narrative and noted on several occasions that the Spanish colonists had a detrimental impact on the indigenous communities. Tylor routinely highlighted instances where the indigenous Mexican people were being subjugated and exploited by their European counterparts. He wrote,

> At the city-gate [of Vera Cruz] stands a sentry – the strangest thing I ever saw in the guise of a soldier – a brown Indian of the coast, dressed in some rags that were a uniform once, shoeless, filthy in the extreme, and armed with an amazing old flint-lock. He is bad enough to look at, in all conscience, and really worse than he looks, for no doubt – he has been pressed into the service against his will, and hates white men and their ways with all his heart. Of course he will run away when he gets a chance; and, though he will be no great loss to service, he will add his mite to the feeling of hatred that has been growing up for these so many years among the brown Indians against the whites and the half-cast Mexicans.[43]

Tylor's travel experience had a significant effect on his ethnological writing for several reasons. Firstly, by travelling abroad and seeing indigenous peoples *in situ*, Tylor could claim an authoritative understanding of ethnological subjects and collect substantive data upon which to base his research claims. Secondly, because he did not study medicine or natural history at a university, as many of his ethnological and anthropological counterparts had, his time in Mexico under the guidance of his mentor Christy can be seen as an intensive practical training course in ethnography. Thirdly, Tylor came to value first-hand observation because it meant that the researcher was not solely reliant on secondary accounts, which potentially misrepresented or underemphasized aspects of a culture.

When he returned to England in the late 1850s, Tylor's status quickly rose within the ethnological community. By the early 1860s he was becoming a leading member at ESL, holding close ties with Huxley's younger circle of scientific naturalists and Hodgkin and Christy's older group of Prichardian monogenists.[44] Tylor even had connections with Hunt and the ASL, serving as the society's foreign secretary in 1863–4 before – according to Stocking – he was enraged by Hunt's 'pugnacious racism' which 'offended his humanitarian Quaker beliefs'.[45] In 1865 Tylor also published his first major ethnological work, entitled *Researches into the Early History of Mankind and the Development of Civilisation* (1865), a title chosen in order to indicate that it was building upon Prichard's earlier work.[46]

Tylor did not join Huxley and John Lubbock in their attack on anthropology in the 1860s, and consequently, because he distanced himself from these debates, he was not a target of Hunt's staunch criticisms of the ethnological community. This could possibly explain why, during the aftermath of the schism in the early 1870s, Tylor was able to continue to build upon his first-rate reputation; his name was not directly associated with either camp. At the beginning of the 1870s, when the newly amalgamated anthropological community was looking for fresh leadership, he was able to acquire a leading role within the Anthropological Institute. In addition, because his family owned a successful brass foundry, Tylor was able to devote himself entirely to the study of human diversity. It was not necessary for him to find gainful employment, because his family's wealth could support his research. This enabled him to produce many anthropological works throughout the second half of the nineteenth century, and he was even appointed the first Reader in Anthropology at Oxford in 1884.

## Developmental Theories and Human Variation

> It seemed to me sufficient to indicate, in the first edition of my 'Origin of Species', that by this work, 'light would be thrown on the origin of man and his history'; and this implies that man must be included with other organic beings in any general conclusion respecting his manner of appearance on this earth.[47]
>
> Charles Darwin, 1871

### *Darwin, Species Debates and Sexual Selection Theory*

Darwin was defensive of his theories in the opening pages of his *Descent of Man*, and much of what he argued in the book responded to many of the debates we examined in previous chapters. During the 1860s, evolution – and specifically natural selection – was routinely criticized by members of both the ethnological and anthropological communities. Even Huxley, who was one of Darwin's closest allies, had some reservations about the mechanism of natural selection, especially the notion that 'nature made no leaps'.[48] Nevertheless, by the 1870s, with Hunt – one of Darwin's staunchest British critics – deceased, and the evolutionist Lubbock presiding over the Anthropological Institute as its president, Darwin believed that the majority of the younger generation of naturalists were gradually accepting his ideas, even if some had reservations about certain components of his theoretical model. He wrote,

> a large number [of naturalists] must admit that species are the modified descendants of other species; and this especially holds good with the younger and rising naturalists. The greater number accept the agency of natural selection; though some urge, whether with justice the future must decide, that I have greatly overrated its importance.[49]

Another chief preoccupation for Darwin was to refine and clarify the older theories and methodologies of researchers interested in human diversity. A main point of contention between Hunt and Huxley during the 1860s had to do with whether Darwinian evolution applied to humans. After all, human variation was not a central focus of Darwin's *Origin of Species*, and he only hinted in the introduction that his theories applied to humanity. Instead, Darwin demonstrated his central thesis by using evidence from both the plant and animal kingdoms.[50] However, in 1871, after nearly twelve years of uncertainty regarding this issue, Darwin finally decided to bring some clarity to these debates. In the introduction of the *Descent of Man*, he outlined for his readers the analytical component of his observational programme and emphasized how his work applied to humans. He wrote, 'The sole object of this work is to consider, firstly, whether man, like every other species, is descended from some pre-existing form; secondly, the manner of his development; and thirdly, the value of the differences between the so-called races of man'.[51]

Under this research model, all organisms within the natural world were affected by evolutionary mechanisms, and humans were like any other natural history specimen. This was significant for two reasons. Firstly, Darwin was attempting to demonstrate how his theories worked in relation to the races of Man, and secondly, he was challenging the traditional view that humans possessed a privileged position within the natural world. Thus *Descent of Man* was fundamentally a natural history work that asked anthropological questions. As such, Darwin devoted as much of his discussion to humans as he did to animals. This approach built upon Huxley's arguments from the 1860s, which urged practitioners to consider humans in relation to their environment, with special attention to plants and animals.[52]

Darwin's larger natural history focus is further apparent when one looks at how he treated traditional ethnological classificatory methodologies in his work. Taxonomies of human varieties were a standard component of early nineteenth-century books on the sciences of Man.[53] However, Darwin distanced himself from this observational practice and stated that it was not 'necessary to describe in detail the differences between the several races' and classify them into groupings based on their physical structure. He believed that such a method had already been outlined thoroughly 'in many valuable works', including Prichard's *Researches*, which – according to Desmond and Moore – was a staple reading for Darwin leading up the publication of his *Origin of Species*.[54] Instead, Darwin's objective was to further transform and refine the analytical practices that researchers used when making sense of their data, by providing them with a sound theoretical basis for substantiating their claims. As Desmond and Moore noted, 'Prichard stockpiled evidence to prove the races were one; Darwin went beyond that to plot hypothetical mechanisms that splinter them, push them apart – in short, point out how they could have *originated naturally*'.[55] It was a much more comprehensive theoretical paradigm based fundamentally on natural history principles.

Darwin also wanted to challenge his opponents, who argued that the current state of scientific theories and methods were incapable of explicating Man's origin because so much time had passed since the first humans inhabited the earth.[56] Darwin rejected this stance and argued that 'It has often and confidently been asserted, that man's origin can never be known: but ignorance more frequently begets confidence than does knowledge: it is those who know little, and not those who know much, who so positively assert that this or that problem will never be solved by science'.[57] As a response, Darwin stressed that his extensive framework would trace Man's progression, showing with substantive evidence how variation occurred. He believed that he was firmly placing the discipline of anthropology on a sound evolutionary model.

The first six chapters of the first volume of *Descent of Man* discussed how humans evolved from animals, tracing their physical, mental and moral developmental processes.[58] The existing form of races was not discussed until chapter seven, entitled 'On the Races of Man'. This chapter is particularly interesting because it demonstrated how Darwin's anthropological ideas built upon many of the earlier studies on human variation from first half of the nineteenth century. Darwin began by stating that his aim was to outline for his reader a systematic programme that showed how it was possible to uncover Man's origin by comparing and analysing the varying physical conformations of different races. He wrote, 'It is not my intention here to describe the several so-called races of men; but to inquire what is the value of the differences between them under a classificatory point of view, and how they have originated'.[59] Instead of describing what a typical representative of the Zulu looked like, as was the case in works such as Latham's *Varieties of Man*, Darwin wanted to explain the causes that shaped their physical appearance. This subtle difference in focus, according to Darwin, was theoretically more sophisticated. Darwin did not simply describe the physical appearance of different races, but explained the reasons for why they possessed certain features.

He continued by discussing several long-standing concerns within ethnological and anthropological circles, examining issues such as whether human variation should be interpreted as difference of 'variety' or 'species'. He wrote,

> In determining whether two or more allied forms ought to be ranked as species or varieties, naturalists are practically guided by the following considerations; namely, the amount of difference between them, and whether such differences relate to few or many points of structure, and whether they are of physiological importance; but especially whether they are constant.[60]

Darwin believed that in order to settle this classificatory debate regarding terminology, practitioners within the sciences of Man should examine the standard conceptual understandings of each word. Thus, for instance, he discussed how naturalists had traditionally conceived the term 'species' and wrote, 'Constancy of character is what is chiefly valued and sought for by naturalists. Whenever it can be shewn, or rendered probable, that the forms in question have remained distinct for a long period, this becomes an argument of much weight in favour of treating them as species'.[61] Under this model, if a characteristic was frequently observed over several generations, the organism under study could be classified as a species. The issue of sterility came to the fore, because naturalists interested in human diversity customarily associated infertility with arguments for difference of species.[62] If an African and a European procreated and produced offspring who were prolific, monogenists such as Prichard argued that the progenitors were examples of two varieties within the same species. If, however, the offspring were sterile, the parents were thought to be of different species.[63] Darwin wrote, 'Even a slight degree of sterility between any two forms when first crossed, or in their offspring, is generally considered as a decisive test of their specific distinctness'.[64] However, he rejected this hypothesis because he argued that sterility was an unreliable test for determining whether two similar organisms were from the same species. According to Darwin, there were many instances in nature where two distinct species produced fertile progeny.[65]

Geographic distribution was also an important point for Darwin, and he believed it warranted further consideration. Without identifying any individuals, he claimed that some researchers used examples of tribes located in remote places – with limited technological prowess for long-distance navigation – to support arguments for difference of species. Under this model, these practitioners argued that it was improbable for indigenous groups such as the Tasmanians and 'civilized societies' such as the English to derive from a common location. If it were the case that they shared an ancestry, how did these 'savage people' circumnavigate the globe to remote islands, given their current navigational technologies? Darwin wrote, 'Geographical distribution is often unconsciously and sometimes consciously brought into play; so that forms living in two widely separated areas, in which most of the other inhabitants are specifically distinct, are themselves usually looked at as distinct'.[66]

A further enduring area of contestation within the sciences of Man was the problem of how to make sense of difference. Using his credentials as a seasoned traveller, Darwin remarked that initially it can be difficult to identify subtle differences among subtypes of the same human variety. Moreover, he added that it was even more difficult to see the commonalities between extreme variations. He wrote, 'If a naturalist ... were to compare a Negro, Hottentot, Australian, or Mongolian, he would at once perceive that they differed in a multitude of characters'.[67] However, by spending time observing human diversity *in situ*, he argued that it was possible to recognize these subtleties and place the various conformations of humans onto a spectrum showing how they all come together.[68] Darwin wrote that 'we must make some allowance for our nice powers of discrimination gained by the long habit of observing [humans variation] ... Even the most distinct races of man ... are much more like each other in form than would at first be supposed'. This was an argument that echoed what Lawrence discussed in his lectures on the *Natural History of Man* from 1819, a text that influenced Darwin's classificatory understanding of races.[69]

Darwin continued by stating that the practice of separating races according to observable structural features was a time-consuming and potentially endless task. He wrote,

> There is, however, no doubt that the various races, when carefully compared and measured, differ much from each other, – as in the texture of the hair, the relative proportions of all parts of the body, the capacity of the lungs, the form and capacity of the skull and even in the convolution of the brain. But it would be an endless task to specify the numerous points of structural difference.[70]

Darwin noted that if one were to observe members from the same tribe, one would identify a tremendous amount of physical variation within the community, and these populations almost certainly shared an ancestry. He stated that 'Savages, even within the limits of the same tribe, are not nearly so uniform in character, as has often been said', and he strengthened this claim by emphasizing his authority as someone who had observed first-hand many different tribal communities during his *Beagle* voyage.[71] Conversely, if researchers were to approach the study of races from an alternative perspective and examine the commonalities between different ethnic groups, they would begin to see the unity of races because certain chief traits were constant in all humans.

Darwin also challenged those practitioners who focused on differences of physical character, by critiquing the method of classifying humans according to cranial features. He wrote, 'The shape of the skull varies much in some races; and so it is with every other character. Now all naturalists have learnt by dearly-bought experience, how rash it is to attempt to define species by the aid of inconstant characters'.[72] This was a response to Hunt, who challenged monogenetic taxonomical systems because they overemphasized similarities in structural form. According to Hunt, more emphasis should be placed on variation. In doing so, it would show that the human races were distinct species.[73]

Darwin responded by stating that many characteristics can vary within any given species because of an abundance of adaptive causes. However, he reiterated that it was those features that remained constant which brought the different races together into a single grouping.[74] Darwin returned to the idea of a racial spectrum that demonstrated how the various gradations observed in racial types came together. He wrote, 'This diversity of judgement does not prove that the races ought not to be ranked as species, but it shews that they graduate into each other, and that it is hardly possible to discover clear distinctive characters between them'.[75] In lieu of this point, Darwin argued that human races should be conceived as 'sub-species' from a single ancestry.[76]

Finally, one of Darwin's most significant contributions to the sciences of Man was his theory of 'sexual selection' because it formulated an extensive conjectural mechanism for explaining both typical and atypical characteristics in different racial populations. Darwin's interest in this selection process and its relation to humans built upon earlier ethnological models. In the first edition of the *Researches* from 1813, Prichard discussed what he called the 'hereditary transmission of the varieties' and linked this to concepts of beauty. This idea about societal conceptions of beauty was further extended in the third edition of the *Researches* published between 1836 and 1847. In essence, Prichard believed that different populations favoured certain physical characteristics – such as blue eyes – over others. Thus over several generations, this favouritism produced archetypal anatomical features within a community.[77]

Other ethnologists such as Latham struggled to explain aberrant characteristics in different ethnic groups. In the *Varieties of Man*, Latham attempted to understand why certain subtypes such as the Papuans – who were part of the Mongole variety – acquired features similar to Africans when there was no evidence to suggest that the two populations shared common ancestry through cross-cultural encounters. He suggested that there was a selective mechanism at work that connected to procreation and various other external influences such as diet and environment. He described these instances as examples of 'deviation with transition'.[78] Darwin's theory of sexual selection expanded upon both these earlier examinations by providing readers with a much more rigorous theoretical explication of the process.[79]

In volume two of the *Descent of Man*, Darwin discussed different aspects of the body and cross-compared the way various races perceived them aesthetically. A quality that was deemed beautiful in one society may not be in another. He wrote that 'As the face with us [the English] is chiefly admired for its beauty, so with savages it is the chief seat of mutilation. In all quarters of the world the septum, and more rarely the wings of the nose are pierced, with rings, sticks, feathers, and other ornaments inserted into the holes'.[80] In both cases, Darwin claimed that the face was a defining feature of beauty. However, in the English example, it was admired for its natural qualities, while for indigenous people it was admired for its decorations. Darwin continued by discussing other examples of how different races conceived beauty and the various factors that influenced their attraction. He stated that 'We have seen ... that with all barbarous races ornaments, dress, and external appearance are highly valued; and that the men judge of the beauty of their women by widely different standards'.[81]

Much of Darwin's anthropological writings from the *Descent of Man* were responses to the criticisms he received during the 1860s. There are many instances where Darwin outlined the problems he saw with his opponents' methodologies. His aim was to demonstrate for his readers how evolutionary mechanisms applied to humans. At the same time, Darwin also tended to shift away from discussions that prioritized humans as being distinct in nature, by emphasizing that human development worked in the same way as it did with the rest of the natural world. Finally, by examining Darwin's work closely, we can see multiple examples of how his observational practice built upon earlier ethnological research from the first half of the nineteenth century. In a sense, Darwin's *Descent of Man* can be positioned as a synthesis of older theories and methods. The difference, however, was that Darwin's theoretical programme – especially in regards to both his natural and sexual selection theories – transformed the way researchers understood human diversity. Not only did it provide practitioners with didactic descriptions of the various races, but it also explained the causes that created human variations.

## *Tylor, Cultural Development and the Concept of Survivals*

Many of the methods and theories outlined in Tylor's *Primitive Culture* built upon and moved beyond earlier techniques. In the preface of the first edition from 1871, Tylor acknowledged that the ideas and methodologies discussed in his book were the result of refining his early analysis from the *Researches into the Early History of Mankind*.[82] Tylor was primarily concerned with demonstrating how non-physical manifestations in different societies developed into more intricate conceptions. For example, how did the belief in spirits animating the natural world evolve into organized religions such as Christianity?[83] Tylor argued that to answer these sorts of questions scientifically, the discipline required a more rigorous analytical programme, with special attention paid to the way researchers made sense of their data. He wrote, 'The statement of the facts must form the staple of the argument' in order to substantiate any claim empirically.[84] The source of his evidence was an important component of his observational practice and it added further credibility to his work. He wrote, 'The authorities for the facts stated in the text are fully specified in the foot-notes, which must also serve as general acknowledgement of obligations to writers on ethnography and kindred sciences, as well as to historians, travellers, and missionaries'.[85]

Tylor's emphasis on facts is suggestive when one considers the anthropological discussions of the 1860s. Both Hunt and Huxley argued for the systematization of ethnographic evidence according to detailed instructions. Furthermore, these figures saw data relating to the anatomy and physiology of different races as a primary concern for the sciences of Man.[86] Tylor's preoccupation with regards to ethnographic facts was a continuation of these practical models, with one major difference: it was chiefly concerned with non-physical attributes. Because cultural features such as norms and values were abstract concepts, Tylor believed that this kind of data required more scrupulous examination before it could be used effectively in an empirical study. He wrote, 'Should it seem to any readers that my attempt to reach this limit sometimes leads to the heaping up of too cumbrous detail, I would point out that the theoretical novelty as well as the practical importance of many of the issues raised, make it most unadvisable to stint them of their full evidence'.[87]

Tylor continued by stating that 'In the course of ten years chiefly spent in these researches, it has been my constant task to select the most instructive ethnological facts from the vast mass on record'.[88] Looking at his journals, it is striking how many sources Tylor analysed in his studies. In total there are fifty-five notebooks containing his observations on books ranging from travelogues and folklore stories to philological tractates. For each book that Tylor read, he would record its author, shorthand title and in some instances the date of publication. He then proceeded by outlining in point form important passages with page numbers, so that he could at a later time examine key excerpts in more detail. For example, when he read John Crawfurd's *Grammar and Dictionary of the Malay Language* (1852),[89] Tylor wrote the following sketch of the first volume of the book:

> xx: Sanskrit words + Javanese grammar.
> xxxiii: Words euphemized in polite language.
> xlii: Indrakila = thunderbolt of Indra, Kingdom of Java.
> xlvi: First records of Malay trade 1st century.
> lviii: One = one *[stone]* – lxvii-ccxviii
> lxxx: Dayak message by *[Nicholas]* Sticks.
> cxviii: Batara = Avatara
> cxxxiv: Limits of Polynesian language.
> clv: Flint = fire stone in Malay & *[Malayasy]*
> cxli: English words in Polynesian & clxxv.
> clviii: French words in *[Mawagasseur]* - *[Professed]*
> clix: Papuans etymology.

cxcix: Rites.
ccix: Words seem *[1 word illeg]* Malay *[*weary*]* from *[*aim*]*.
ccxl: Domestic animals of Polynesia.
ccxlii: *[*Cullimated*]* plants.
ccl: Bowe & cclxxi.
cclxxv: Malay influences in Malaynesean.
ccxci: Malay words adopted in Australia.[90]

Looking at Tylor's notes on Crawfurd's text is instructive because we see how he conducted his observational practice. A close examination of Tylor's journal shows us how he went about collecting his ethnographic information and what sorts of material he was interested in analysing. Tylor was a monogenist and was interested in locating evidence that illustrated a common link between different races. Thus he noted in his journal instances where Crawfurd identified examples of Polynesian languages possessing English and French words in an effort trace a common root between these dialects.

Tylor also contributed to the development of the discipline's lexicon. In the introduction to *Primitive Culture*, he began by coining the term 'culture'. This was important because his book aimed to trace the progression of 'primitive cultures' from 'simple states', including examples such as Maori villages, to more 'complex societies' such as European nations. He wrote, 'Culture or Civilization taken in its wide ethnographic sense, is that complex whole which includes knowledge, belief, art, morals, law, customs, and any other capabilities and habits acquired by man as a member of society'.[91] Under this model, cultural studies examined not only social organization, but norms and values as well. The aim of an ethnographer was to bring together a vast array of data and show how any given culture functioned. Tylor also argued that by cross-comparing different 'lower tribes' and 'higher nations', it was possible to show the uniformity of races and demonstrate the evolutionary processes that all societies undergo.[92] He stated that 'the uniformity which so largely pervades civilization may be ascribed, in great measure to the uniform action of uniform causes, while ... its various grades may be regarded as stages of development or evolution'.[93]

Tylor continued by stating that a central problem for studying culture as a scientific subject – as opposed to the anatomical and physiological conformations of races – was that 'The world at large is scarcely prepared to accept the general study of human life as a branch of natural science'.[94] He wrote, 'Too many educated minds there seems something presumptuous and repulsive in the view that the history of mankind is part and parcel of the history of nature'.[95] Tylor argued, however, that by accumulating copious amounts of facts and closely examining this material in detail, it was increasingly possible to reject this stance and make the study of cultures scientific. He stated, 'There are many who would willingly accept a science [of culture] ... if placed before them with substantial definiteness of principle and evidence, but who not unreasonably reject the systems offered to them, as falling far short of a scientific standard'.[96]

Tylor also wanted to ascribe natural laws to cultural development by looking for the causes that shaped societies. It was similar to how Darwin examined the causes that influenced the development of organisms in nature, which he discussed in his *Origin of Species*.[97] Tylor wrote, 'let us take this admitted existence of natural causes and effect as our standing-ground, and travel on it so far as it will bear us. It is on this same basis that physical science pursues, with ever-increasing success, its quest of laws of nature'.[98] However, in order to establish these laws, Tylor discussed the daunting task of making sense of cultural data. He argued that the first problem for researchers was to deal with the 'enormous complexity of evidence, and imperfection of methods of observation'.[99] In order to deal with these issues, the first step for Tylor was to establish a more manageable system for working with ethnographic data.

This was nothing new for the sciences of Man, and it built upon earlier ethnological observational programmes such as Prichard's taxonomical system from the first edition of the *Researches* and Huxley's 'well-considered-plan' from the late 1860s.[100] Tylor believed that by dividing cultural traits into groupings such as religion, law and art, it was easier for practitioners to trace the development of these attributes from simple concepts to more sophisticated ones. He stated that 'the phenomena of Culture may be classified and arranged, stage by stage, in a probable order of evolution'.[101] By showing how all cultures developed through stages, the ultimate aim of Tylor's work was to demonstrate the unity of humans. He wrote, 'Surveyed in a broad view, the character and habit of mankind at once display that similarity and consistency of phenomena which led the Italian proverb-maker to declare that "all the world is one country", "tutto il mondo è paese"'.[102]

As was the case in Darwin's *Descent of Man*, Tylor argued that it was those features that were observed routinely in different cultures that brought the various races together.[103] Therefore anthropologists were to focus on the commonalities among different populations, and Tylor wrote, 'this similarity and consistency may no doubt be traced, and they may be studied with especial fitness in comparing races near the same grade of civilization'.[104] These similarities manifested themselves in all forms of cultural aspects, and, for example, he wrote, 'Examine for instance the edged and pointed instruments in such a collection and see how they repeat themselves with wonderful uniformity in the museum shelves which illustrate the life of the lower races'.[105]

If you contrast this statement with the way in which objects are displayed at the Pitt Rivers Museum today, we can see Tylor's lasting influence on the exhibits. After all, he played a central role in establishing its ethnographic collections.[106] In the main showroom of the museum, comparable objects are placed together into showcases to emphasize the unity of cultures. If you look at the display case marked 'Drums', you will see examples of similar-looking percussion instruments – crafted by different indigenous communities – placed next to each other in order to emphasize their shared characteristics.[107] Tylor's use of ethnographic displays is further significant because it built upon earlier ethnological training sites, such as Latham's exhibit from the Crystal Palace in the early 1850s. This link between Tylor and Latham is further strengthened because some of the only remaining photographs of the ethnological department at Sydenham come from Tylor's personal collection. Feasibly, the reason he held on to these images was because he viewed Latham's exhibition as an important ethnological site. The Pitt Rivers Museum can also be positioned as a training site because researchers were able to refine their observational skills by examining Tylor's carefully arranged cabinets. It is another example of how he built upon and transformed earlier techniques.[108]

As was the case in Darwin's *Descent of Man*, Tylor also wanted to outline for his readers a new kind of taxonomy that was not simply based on organizing different ethnicities into groups according to their anatomical and physiological characteristics. Tylor did not see the value of producing yet another classification of this type because such a method had already been organized and refined several times in many earlier ethnological works. Instead, moving beyond these pre-existing systems, Tylor wanted his readers to work under the assumption that all humans were fundamentally homogenous in nature and that their external features were not essential for a cultural model. He stated that

> For the present purpose it appears both possible and desirable to eliminate considerations of hereditary varieties or races of man, and to treat mankind as homogeneous in nature, though placed in different grades of civilization ... The details of the enquiry will, I think prove that the stages of culture may be compared without taking into account ... bodily configuration and the colour of their skin and hair.[109]

Thus Tylor outlined a modified version of a traditional classificatory system – organized by figures such as Latham in his *Varieties of Man* – by redefining the taxonomic method along cultural lines. The ultimate aim was to naturalize non-physical attributes and strengthen their empirical value for science. He wrote,

> A first step in the study of civilization is to dissect it into details, and to classify these in their proper groups. Thus, in examining weapons, they are to be classed under spear, club, sling, bow and arrow, and so forth; among textile arts are to ranged matting, netting, and several grades of making and weaving threads ...[110]

Tylor proceeded by explicating how ethnographers were to go about transforming cultural objects from different societies into natural history specimens. Stocking observed that *Primitive Culture* 'was an attempt to treat the old ethnological problem of the unity of human species in a radically lengthened temporal context ...

[and] account for the growth of culture in natural developmental terms'.[111] Tylor explained that 'To the ethnographer, the bow and arrow is a species, the habit of flattening skulls is a species, the practice of reckoning numbers by ten is a species'.[112] He continued by emphasizing that 'Just as the catalogue of all the species of plant and animals of a district represents its Flora and Fauna, so the list of all the items of the general life of a people represents that whole which we call its culture'.[113]

Another central issue for Tylor had to do with the quality and reliability of the ethnographic reports informants were collecting during their travels abroad. Tylor argued that the ethnographer had to use their judgement when using quotes from colonial officers, travellers and missionaries because these accounts often misrepresented the indigenous people described in them. He wrote,

> This question [of credibility] is, indeed, one which every ethnographer ought to keep clearly and constantly before his mind. Of course he is bound to use his best judgement as to the trustworthiness of all authors he quotes, and if possible to obtain several accounts to certify each point in each locality.[114]

This was an extension of earlier ethnological techniques. Practitioners such as Prichard and Lawrence were concerned about the frequency with which informants' travel reports embellished their cross-cultural encounters with indigenous peoples. To limit the amount of discrepancies in future works, both Prichard and Lawrence explicated for their readers why it was important to compare multiple accounts before attributing a characteristic to a race.[115]

However, Tylor did not want his reader to lose sight of the ethnographer's primary aim, which was to demonstrate the unity of humans. He argued that the role of a researcher, when comparing different members of a community, was to examine closely those commonly observed traits and bring these similarities to the fore of the analysis. He wrote,

> There is found to be such regularity in the composition of societies of men, that we can drop individual differences out of sight, and thus we can generalize on the arts and opinions of whole nations, just as, when looking down upon an army from a hill, we forget the individual soldier, whom, in fact, we can scarce distinguish in the mass, while we see each regiment as an organized body, spreading or concentrating, moving in advance or in retreat.[116]

By focusing on similitude, the data would demonstrate the singularity of humankind and allow the researcher to claim individual difference as insignificant.

Tylor developed several other analytical techniques for tracing cultural development. He wrote, 'Progress, degradation, survival, revival, modification, are all modes of connexion that binds together the complex network of civilization'.[117] However, a discussion of all these modes is far too encompassing for a discussion here. As such, let us examine one of these methods and consider his concept of 'survivals'. Tylor described survivals as follows:

> Among the evidence aiding us to trace the course which the civilization of the world has actually followed, is that great class of facts to denote which I have found it convenient to introduce the term 'survivals'. These are processes, customs, opinions, and so forth, which have been carried on by force of habit into a new state of society different from that in which they had their original home, and they thus remain as proofs and examples of an older condition of culture out of which a newer has been evolved.[118]

For instance, proverbs were a type of survival, and Tylor claimed that they represented antiquated understandings of the natural world. They were useful sources of data for his interpretive method because they enabled researchers to trace changes in the cultural explanatory models of different communities over hundreds of years. Tylor stated that

> On the strength of these survivals, it becomes possible to declare that the civilization of the people they are observed among must have been derived from an earlier state, in which the proper home and meaning of these things are to be found; and thus collections of such facts are to be worked as mines of historic knowledge.[119]

Tylor also discussed 'partial survivals', which were old habits or objects that were adapted into new circumstances. He used the example of children playing with bows and arrows as an instance of partial survivals because it was a type of outdated technology that continued to exist within the society despite newer devices such as guns replacing them.[120]

Tylor's cultural developmental theory charted the various stages societies underwent as they evolved from 'simple tribes' to 'complex nations'. By tracing this process, he developed a rigorous observational programme for working with cultural data that responded to researchers such as Hunt and Huxley, who during the 1860s viewed non-physical characteristics as less empirically grounded types of evidence for substantiating scientific claims. Moreover, Tylor outlined an analytical method for naturalizing cultural attributes such as values, customs and norms by transforming them into natural history specimens. Tylor's framework built upon and re-configured Prichardian methods, which attempted to establish the unity of races by tracing the non-physical pronouncements of tribes and nations to a single source. In many respects Tylor can be positioned as a synthesizer of ethnological and anthropological practices from the first half of the nineteenth century.

## More Instructions for Informants

> It is the duty of every civilized traveller in countries newly opened up to research to collect facts, pure unvarnished facts for the information of those leading minds of the age, who by dint of great experience, can ably generalize from the details contributed from diverse sources.[121]
>
> Harry Johnston, 1889

### *Darwin's Questions on Emotions*

In 1872 Darwin published another book on the study of humans, entitled *The Expression of Emotions in Man and Animals*.[122] As the title suggests, Darwin aimed to show how humans and animals expressed emotions and – as was the case in *Descent of Man* – it was fundamentally a natural history text that asked anthropological questions because it did not privilege humans as the sole object of its study. However, the focus of this section is not on Darwin's *Expression of Emotions*, but rather on his instructions for informants that formed the basis of his analysis in this book. With these concerns in mind, it will examine his questions on emotions and some of the responses he received from his correspondents living abroad.

The importance of Darwin's correspondence network was a major theme in Janet Browne's book *Charles Darwin: The Power of Place*. In her discussion, she argued that Darwin's location in Kent redefined his scientific practice. His location shaped his daily activities in two ways. Firstly, by not being in London, Darwin was reliant on his correspondents for news on the scientific, religious and social debates occurring in the British capital at places such as the Royal Society, the Royal Institution, the ESL and the ASL. Secondly, Darwin substantiated much of his theories with data collected by informants living throughout the world. Thus he was reliant on a large network of communication for his research. Letter writing was an important part of his daily routine. Browne writes, 'His study inside Down House became an intellectual administration and calculation, in which he churned out requests for information and processed answers, kept himself at the leading edge of contemporary science, and ultimately orchestrated a transformation of Victorian thought'.[123]

Looking at Darwin's engagement with his informants is instructive because it demonstrates how he sought to improve the analytical component of his observational practice by acquiring a higher quality of data. Direct examination was fundamental for Darwin, and because thirty years had passed since he travelled on board the *Beagle*, it was important for him to collect fresh evidence and not rely on his own memory, which he recognized was affected by the passage of time. It was not the first time Darwin was involved in a project that directed the collecting practices of informants living abroad. Darwin was one of the figures involved in writing the *Queries Respecting the Human Race to be Addressed to Travellers and Others* from 1839.[124]

In the introduction of *Expression of Emotions*, he wrote, 'I circulated, early in the year 1867, the following printed queries with a request ... that actual observations, and not memory, might be trusted'.[125] However, Darwin noted that he was unhappy with the questions he sent to informants because they 'were written after a considerable interval of time, during which my attention had been otherwise directed, and I can now see that they might have been greatly improved'.[126] Questionnaires such as Hodgkin and Cull's ethnological manual from the early 1850s were carefully designed to ask questions that any informant could collect regardless of their training. For example, they wrote, 'What is the ordinary size of families, and are there any large ones?'[127] Darwin's questions, however, were much more specialized, and because they requested data relating to bodily reactions to emotions, the information was difficult to collect.

One of Darwin's primary aims was to determine whether all races expressed emotions in a similar way. Thus he sent his instructions to various correspondents who were living in different regions of the world and who had some familiarities with the indigenous peoples. In an 1867 letter to the naturalist Johann Friedrich Müller (1822–97), who was living in Brazil, Darwin wrote, 'It occurred to me that you might without much trouble make a few observations for me in the course of some months on Negroes, or possibly S. Americans'.[128] He included a supplementary enclosure in the letter with the following seventeen questions.

(1) Is astonishment expressed by the eyes & mouth being opened wide & by the eyebrows being elevated?
(2) Does shame excite a blush, when the colour of the skin allows it to be visible?
(3) When a man is indignant or defiant does he frown, hold his body & head erect, square his shoulders & clench his fists?
(4) When considering deeply on any subject or trying to understand any puzzle does he frown, or wrinkle the skin beneath the lower eyelids?
(5) When in low spirits are the corners of the mouth depressed, & the inner corner or angle of the eyebrows raised by that muscle which the French call the 'grief' muscle?
(6) When in good spirits do the eyes sparkle with the skin round & under them a little wrinkled & with the mouth a little extended?
(7) When a man sneers or snarls at another is the corner of the upper lip over the canine teeth raised on the side facing the man whom he addresses?
(8) Can a dogged or obstinate expression be recognized, which is chiefly shewn by the mouth being firmly closed, a lowering brow & slight frown?
(9) Is contempt expressed by a slight protrusion of the lips & turning up of the nose with a slight expiration?
(10) Is disgust shewn by everted lower lip, slightly raised upper lip with sudden expiration something like incipient vomiting?
(11) Is extreme fear expressed in the same general manner as with Europeans?

(12) Is laughter ever carried to such an extreme as bring tears into the eyes?
(13) When a man wishes to shew that he cannot prevent something being done, or cannot himself do something does he shrug his shoulders, turn inwards his elbows, extend outwards his hands & open the palms?
(14) Do the children when sulky pout, or greatly protrude their lips?
(15) Can guilty or sly or jealous expressions be recognized? Tho' I know not how these can be defined.
(16) As a sign to keep silent is a gentle hiss uttered?
(17) Is the head nodded vertically in affirmation, & shaken laterally in negation?[129]

Darwin added that 'Observations on natives who have had little communication with Europeans would be ... most valuable' because they were conceivably more accurate in representing the way different races expressed emotions, as they were potentially unaffected by European enculturation.[130] He wrote that he wanted detailed responses to his queries because 'General remarks on expression are of comparatively little value'. He argued that 'A definite description ... w$^{d}$ possess much more value, & an answer to any one of the foregoing questions w$^{d}$ be gratefully accepted'. Following the questions would also assist in standardizing the data because each informant would collect material according to the same set of instructions.[131]

By the end of 1867, Darwin began receiving letters with answers to his instructions on emotions. James Weale, who was a naturalist living in Port Elizabeth in South Africa, sent responses to Darwin's list of questions and wrote that 'Kaffirs when in grief place their hands especially the women on their heads. They likewise do so to express surprise. They also place one hand on the chin & mouth in grief'.[132] Once Darwin correlated the information from his correspondents, he proceeded by analysing the data. This evidence formed the basis of his *Expression of Emotions*. However, an analysis of this book would be far too comprehensive for an examination here.

Darwin was highly attuned to the problems associated with armchair studies, and he recognized the importance of using data based on direct observation when putting together his work on the *Expression of Emotions*. To limit the amount of discrepancies in his work, one of his primary aims was to raise the quality and standard of the material collected by informants living abroad. In doing so, he hoped that his correspondents would send similar types of evidence – because they all used the same questions – which would enable him to conduct exacting cross-comparative analyses. However, Darwin was not the only researcher developing more specialized ethnographic questionnaires in the late 1860s and early 1870s. From 1871 onwards, a group of figures at the Anthropological Institute were also developing instructive literature.

## *The Anthropological Institute's First Questionnaire*

At the beginning of the 1870s the newly amalgamated British anthropological community began organizing several different projects aimed at transforming the quality of the discipline's research materials. One of these initiatives was to produce a revised ethnographic questionnaire that incorporated many of the new methodologies organized during the 1860s. The existing instructive texts were beginning to look outdated because neither the ESL nor the ASL had published a questionnaire since Hodgkin and Richard Cull's manual from the early 1850s.[133] With these concerns in mind, in 1872 several leading members of the Anthropological Institute – including Tylor – agreed to work collaboratively and write a guidebook initially designed for an unspecified voyage of exploration to the Arctic, with questions, however, carefully designed so that they were applicable for ethnographic observation in any region of the world.[134]

In 1873 the Anthropological Institute published this questionnaire – entitled 'Questions for Explorers (With Special Reference to Arctic Exploration)' – in its journal. The questions were divided thematically into eight main sections, with each part written by a different figure specializing in that research area. Section one, entitled 'General Questions', was designed by the craniologist and physician Joseph Barnard Davis (1801–81) and outlined twelve main points of consideration for explorers to record when observing different ethnic groups *in situ*. It focused on traditional ethnological questions that reflected the interests of the older generation of practitioners, of which Davis was a member. Using his training in anatomy and physiology, he requested descriptions of skin, eye and hair colour, the average height and weight of the inhabitants, and detailed accounts of their skulls, with special mention of any visible deformities.[135] He also stated that when possible, the explorer should collect 'as perfect as possible [crania], never leaving anything behind, particularly not lower jaws and teeth. On collection, they should be at once marked with tribal name, *in ink* if possible, to prevent confusion'.[136] Davis also requested other types of information regarding language and clothing, which were traditional areas of ethnographic interest.

Section two was written by Tylor and focused on questions relating to mythology, religion and social organization. It utilized techniques such as cultural survivals and the theory of animism, which were outlined in his book *Primitive Culture*. For instance, Tylor included questions such as 'What ideas have they as to souls and other spirits? What do they think of dreams and visions? are they appearances of spirits? Are trances, etc., set down to [the] exit of soul? Are hysterics, convulsions, etc., ascribed to demoniacal possession?'[137] Tylor also wanted information about morals and values and wrote,

> What actions and dispositions are considered good and bad, virtuous and vicious? Does public opinion make much difference in treatment of virtuous and vicious? Are

> there any set laws and penalties? What restraint is there on theft, murder, adultery, etc.? Do acts count as criminal differently when done on a member of the tribe or foreigners? What is the native law or custom as to vengeance? What are the laws or customs as to marriage, inheritance, and clanship?[138]

Tylor's section of the questionnaire was specifically designed for collecting data that connected to his cultural approach to studying human diversity. It was a more updated method for acquiring evidence, using the latest anthropological techniques, which were especially suited for researchers interested in conducting Tylorian cultural studies. If we compare his instructions with Darwin's queries on emotions, we can see that Tylor's questions were much more accessible for informants and carefully written so that anyone could feasibly acquire answers to them. Darwin selected his informants from a list of contacts he had available, and he was able to provide each individual with personalized instructions to complement his list of questions. In contrast, the questionnaire produced by the Anthropological Institute was addressed to all travellers in foreign lands. The idea was that once the data was sent back to skilled anthropologists, they would be able to identify the problematic material – and remove it from the data set – and locate the valuable information, which they would include in their analyses.

'Questions for Explorers' also had instructions that were valuable for other groups and not just anthropologists. Augustus Henry Lane Fox (later known as Pitt-Rivers) was both an active member of the Anthropological Institute and a colonel in the British military. He wrote section four on 'Customs Relating to War', which was useful for anthropological researchers as well as the colonial officials. This was significant because the sciences of Man recognized the value of working collaboratively with the government.[139] Anthropologists gained access to a large network of informants, and colonial officials received detailed information on the peoples inhabiting their territories.[140] Thus the questions designed by Lane Fox considered issues such as:

> What are their war weapons [and] are the same weapons used in war and the chase? What is the exact nature of their defensive armour, especially that described as being made of pieces of wood fastened together? Is the throwing stick used in war? What is the accuracy, range, and penetration of a lance projected by this means?[141]

The manual also outlined questions that built upon the anatomical and physiological interest of ethnologists and anthropologists from the 1860s. For example, the anatomist and physician John Beddoe – who replaced Hunt as president of the ASL after his death in 1869 – wrote section six. His exacting and anatomy-focused instructions outlined for readers a collecting method that aimed to maximize the quality of informants' observations. It was divided into three main parts and contained similar instructions to those outlined by Huxley in his 'well-considered-plan', which was sent to the Colonial Office of London in 1869. Beddoe wrote in part A:

> A. The following measurements should be obtained from as many adults of the two sexes as possible.
> 1. Stature: best gotten by means of a graduated rod, in erect posture. Mention whether shoes are worn, and of what thickness.
> 2. Greatest length of head, from the eminence between the eye- brows; with index or other callipers.
> 3. Greatest breadth of head, wherever found, with callipers.
> 4. Greatest breadth of zygomata, also with callipers.
> 5. Span – i.e., distance between tips of middle fingers, arms being expanded.
> 6. Circumference of chest at nipple (in men).
> 7. Ditto after full expansion by forced inspiration (in men).
> 8. Circumference of thigh at fork.
> 9. Distance from fork to ground.
> 1, 6, and 9, are most important.[142]

It is interesting that 'Questions for Explorers' did not include any detailed guidelines on how to visually depict humans. Photographs and drawings were only mentioned in passing.[143] There are several possible reasons for why these instructions were omitted from the questionnaire. For instance, it is conceivable that the text's contributors saw Huxley's and Lamprey's instructions for photographers from the late 1860s as sufficient, and that many anthropologists and informants would have already been aware of these guides. Conversely, perhaps they chose to exclude photographic instructions for practical reasons. For example, it is feasible that many of their informants had limited (if any) access to cameras.

The lack of photographic instructions aside, it is clear that 'Questions for Explorers' combined different methodological approaches to the study of humans – from early ethnological techniques at the beginning of the century, to anthropological methods devised by figures from the ASL, to the latest cultural developmental programmes by figures such as Tylor. As such, it represented a synthesis of the practices researchers were utilizing throughout the first half of the nineteenth century. As the 1870s wore on, the anthropological community continued to work collaboratively and produce more extensive questionnaires for informants to use when observing humans abroad. By 1874 the Anthropological Institute published the first edition of *Notes and Queries on Anthropology*, and it became the main ethnographic questionnaire in the Anglo-American world right up until the 1950s. The Anthropological Institute continually reworked the text over the course of nearly seventy years. As an instructional guide, it taught researchers and travellers how to record observations *in situ*, and it formed the basis of many anthropological studies in the second half of the nineteenth century.

## Conclusion: Evolutionism and Beyond

In the preface of the second edition of *Primitive Culture* from 1889, Tylor remarked, 'It may have struck some readers as an omission, that in a work on civilization insisting so strenuously on a theory of development or evolution, mention should scarcely have been made of Mr. Darwin ... whose influence on the whole course of modern thought on such subjects should not be left without formal recognition'.[144] He continued by stating that 'This absence of particular reference is accounted for by the present work, arranged on its own lines, coming scarcely into contact of detail with the previous works of [this] eminent philosopher'.[145] Perhaps in light of this statement, we should begin to reassess the way in which we have characterized anthropology in the 1870s and 1880s.

Instead of demarcating these decades as the 'evolutionist era' and positioning it as a disjunctive moment in the history of the discipline, it is more instructive to view the period as part of the continuing attempt by researchers within the sciences of Man to transform their theories and methodologies.[146] What we see at the start of the 1870s was not a new form of anthropology, but rather a synthesis of practices that was part of the continual transformation of the discipline's theoretical models. These shifts in disciplinary practices further refined the older observational methods utilized by ethnologists and anthropologists during the first half of the nineteenth century. By bringing together the research activities of earlier practitioners, Darwin and Tylor changed the discipline by outlining two complementing theoretical frameworks for understanding the processes that all humans undergo as they develop from 'savage states' into more 'civilized nations'.

During the 1880s the cultural approach to studying human diversity increasingly dominated the research field, while Darwin's influence on the discipline waned. Figures such as Tylor continued to develop new types of instructive literature for researchers and informants, as well as expand upon the analytical techniques anthropologists were to utilize when making sense of their data. In 1881 Tylor wrote the first anthropological textbook, which provided novice anthropologists with a set material designed to teach new researchers how to observe human diversity in a specialized way.[147] In 1888 he also published an article 'On a Method of Investigating the Development of Institutions; Applied to Laws of Marriage and Descent', which outlined a theory that explicated an observational technique for interpreting social organizations among different societies.[148] Other researchers built upon Tylor's cultural model, and from the 1870s onwards, figures such as Edward Clodd (1840–1930) published significant folklore studies such as his book *The Childhood of Religions* (1887). The foundation of this work built upon Tylorian methodologies. During the same period, the anthropologist James Frazer (1854–1941) also recognized his debt to Tylor's writings and based his theories of 'totems' from *Totemism* (1887) and *The Golden Bough* (1890) on the principles outlined in *Primitive Culture*.[149] Thus the younger generation of cultural anthropologists were greatly influenced by Tylorian anthropology.

# CONCLUSION

> Perhaps some day [anthropologists] ... will awake to the fact that the time spent in these armchair studies could be much more profitably occupied. Only natives can explain the meaning of the majority of designs and patterns, and not even all of them are sure as to what was meant by the artist in every case.[1]
>
> Alfred Cort Haddon, 1909

## Disciplinary Transformations after 1871

This book has argued against the misconception that early nineteenth-century anthropology was fundamentally an armchair pursuit, with minimal analytical reflection, and detached from the activities of informants who were collecting and recording data in the field. This vision of the discipline's past started to take shape after Darwin and Tylor's heyday. During the 1890s the younger generation of naturalists writing on human diversity topics began to challenge the techniques of the older generation. One of the biggest critics of pre-1890s human variation studies was the naturalist Alfred Cort Haddon (1855–1940). Initially trained in biology and zoology, Haddon became an authority within anthropology after participating in the Cambridge Expedition to the Torres Straits (1898–1901). This expedition had a major effect on the observational practices researchers utilized in their studies of humans. In particular, it encouraged the use of new forms of technology such as sound recording and film to further enrich anthropological observational techniques.[2]

The Torres Straits Expedition was also significant because it was recognized as one of the first university-sponsored ethnographic projects to use a team of researchers to collaboratively collect evidence *in situ*.[3] The *Times* reported on 10 March 1898 that individual '[t]ravelling scholars are often sent out by Oxford and Cambridge but this is probably the first time that a real exploring expedition has been sent out by either University'.[4] The Cambridge Expedition was also canonized as a key moment for the discipline by members of the British anthropological community. Haddon's friend and obituarist Henry John Fleure (1877–1969) wrote that 'The Torres Straits expedition was epoch-making for the science of man' and continued by stating that under Haddon's leadership, the organizers of the expedition selected colleagues from 'cognate branches of study ... so as to make possible a comprehensive integration of their observations'.[5] By bringing together researchers from allied disciplines such as psychology, botany and zoology, the expedition aimed to collect as many types of data as possible on the island and its people. In doing so, it would provide researchers with an extensive understanding of the cultures under observation, which in turn would produce rich accounts of the islanders.[6]

It was also through this experience of spending an extended period of time living among extra-Europeans in their natural environment that Haddon became critical of earlier Victorian anthropology because of its reliance on second-hand reports from informant networks. He wrote that scientists without direct field experience were simply 'retailing second-hand goods over the counter'.[7] Haddon was critical of the collecting methods utilized by travellers, and he believed that a key problem was that informants who provided data for researchers back in Britain often only spent short periods of time observing indigenous populations *in situ*. Therefore, they had limited opportunities to collect substantive evidence. This criticism was reflected in many of Haddon's books and reviews. He wrote that 'As the investigators travelled over such an extensive area and often had but a limited time at the places which they visited, the original observations must be considered as being more of the nature of a survey than as a detailed study'.[8] Since Haddon was critical of the way in which travellers collected their data, he believed that 'armchair studies' were inherently flawed because they were based on unreliable evidence. Consequently he put great emphasis on field studies. Haddon was not the only practitioner to attack earlier Victorian anthropological practices as antiquated. His contemporaries W. H. Rivers (1864–1922) and C. H. Read (1857–1929) were equally concerned about anthropological methodologies. Haddon's criticisms coincided with changes that were occurring within British academia at the beginning of the twentieth century.[9]

In the introduction to this book, we saw that disciplines in the early twentieth century such as anthropology were trying to establish firm positions for their subjects within universities. Academic institutions recognized the significance of funding lifelong careers for scholars, and as Kuklick has put it, 'university reformers undertook to professionalize scholarship, determining that the advance of knowledge required that individuals be able to make lifelong, remunerated careers developing their specialized expertise – just as independent professionals did'.[10] Practitioners such as Haddon – and Bronislaw Malinowski in the 1910s – were keen to further transform the empirical validity of their studies in order to acquire university funding for their research. They argued that a direct engagement with their object of study through fieldwork added to the scientific validity of anthropology's research practices. It was also during this period that Malinowski developed – while observing the Trobriand islanders *in situ* – his self-proclaimed method of 'participant observation', in which the researcher aimed to understand the world through the eyes of the local culture under study by engaging in the rituals, customs and other social activities of the inhabitants. It was one of Malinowski's most significant and lasting contributions to anthropology.

Transformations in tropical medicine also made extended research trips easier because it lowered the chances of practitioners contracting illnesses such as malaria.[11] Early twentieth-century anthropologists recognized the significance of these new medications. Malinowski, for example, referred to his medicine chest in his diary for 1915.[12] The academic calendar also afforded anthropologists working at universities with an opportunity to conduct research during term breaks.[13] A major reason ethnologists and anthropologists in the first half of the nineteenth century relied on informant networks was because they were unable to travel abroad and collect their own data. In many cases, figures such as Prichard and Hodgkin earned a living by practising medicine. Ethnology was something done in their spare time because their medical careers tied them to hospitals and asylums in Britain. However, with new academic posts opening up at universities at the end of the nineteenth century, the younger generations of scholars had more opportunities to conduct their own fieldwork.

The Oxford-trained Baldwin Spencer – who was Professor of Biology at the University of Melbourne – was one of the first scholars to take advantage of these university vacation periods. During the summers of both 1896 and 1897 he was able to conduct some fieldwork. However, Spencer was still partially reliant on informants, and he worked collaboratively with Francis James Gillen (1855–1912), who was an Australian civil servant with a comprehensive knowledge of the Australian aborigines.[14] Spencer is also an interesting example in that he demonstrates how many British scholars forged their scientific careers abroad because of the limited opportunities available at home.[15] This in turn influenced anthropological research because many of these new posts were located in regions where aboriginal populations lived. This made it easier for scholars to conduct ethnographic research in their spare time.

Another tactic that early twentieth-century anthropologists utilized as part of their professionalizing strategies – with its emphasis on 'field-centred research' – was to organize new institutionalized training regimes at universities. Through these initiatives, students learned how to observe, analyse and represent their data according to the guiding principles of leading anthropologists with fieldwork experience. For instance, the London School of Economics became Malinowski's base of operations for a decade, and a number of eminent anthropologists trained under his research programme.[16] It was under this new institutionalized scheme that 'fieldwork' was given a new status, becoming a 'rite of passage' which all anthropologists were to undertake.[17]

Malinowski also envisaged a professionalized discipline of anthropology that worked cooperatively with Britain's imperial administration. He wanted to weed out untrained colonial informants in favour of university-educated anthropologists. Under this scheme, it was important to emphasize that anthropologists possessed important and unique skills that could help colonial officials develop imperial settlements.[18] An enlarged military presence within the imperial world helped to facilitate long-term research trips because it became easier for researchers to reside in colonial settlements for extended periods and gain access to indigenous populations. Transformations in sea, road and rail transport around the world made it possible to reach isolated communities easier and faster.[19] For example, the construction of the Suez Canal – which was completed in 1869 – accelerated the travel time between India and England. Depending on which region of India a ship was sailing to, the travel distance could be cut down by up to 51 per cent.[20]

These institutional, imperial, medical and technological transformations redefined the vocational and disciplinary strategies of late nineteenth-century anthropological researchers. By the early twentieth century, anthropologists juxtaposed their studies as more theoretically informed and accurate in representing indigenous peoples and their cultures. In contrast, they characterized their earlier Victorian counterparts as theoretically naive and methodologically unsophisticated.[21] Victorian anthropology was further discredited during the post-colonial period of the 1960s and 1970s when scholars argued that the discipline had acted as the 'handmaiden of the empire' by providing colonial officials with detailed descriptions of indigenous populations, which in many cases had a devastating impact on extra-European peoples. Consequently anthropologists entered into a period of heightened reflexivity, and they considered the damages the discipline had inflicted upon native peoples. They also became further critical of nineteenth-century anthropology and argued that their Victorian counterparts allegedly lacked first-hand experience with indigenous populations. Thus anthropologists in the 1960s and 1970s framed nineteenth-century research as being objectifying and de-humanizing towards its subjects.[22]

Although historians of anthropology have long since abandoned the notion that nineteenth-century studies of human variation were simply armchair pursuits, there is still much to be gained by asking more insistently what the practices of Victorian researchers entailed. It was the purpose of this book to continue the job of replacing this old polemical dichotomy – between the armchair and the field – with a new set of analytical categories more reflective of the whole activity carried out in the Victorian era. What has been lost above all is the central role of observation for earlier practitioners. Once observational practices come to the fore, the history of anthropology looks far more like one of gradual change due to the adoption of new observational techniques into its methodologies. Field-based research was intended to enhance the quality of anthropologists' observations and should thus be conceived as a continuation of earlier disciplinary practices. This is significant because by reconceiving the discipline in this way, we have a much more fluid picture of anthropology that is not divided into distinct methodological epochs.

## Some Concluding Reflections: Four Major Themes

> No matter what their professional training and special interests, historians inevitably ask questions that are important in our age. They know that past concerns were different from our own, but they must also know how contemporary practitioners view their enterprise; the past may appear different in the future, but knowledge of a discipline's present has some bearing on understanding its history.[23]
>
> Henrika Kuklick, 2008

### *Training Regimes*

There have been several major themes discussed throughout this book, and for our purposes I would like to focus on four of them. The first one is training, and the kinds of instructional preparations that Victorian ethnologists and anthropologists undertook directly influenced their observational practices. As I discussed in the introduction to this book, when I use the term 'observational practices', I am not simply discussing the physical act of looking at an object, but rather something more nuanced. All forms of observation involve laborious training, and researchers from any discipline learn discriminating techniques that seek to identify those characteristics which are of value to the research field.[24] A background in natural history and medicine was particularly useful for studying human diversity during the first two-thirds of the nineteenth century. Anatomical and physiological topics featured prominently in the works of figures such as Prichard, Lawrence, Latham and Huxley. Each of these practitioners emphasized the value of craniological studies for the discipline. Moreover, in all of their works, the skin pigmentation colour of different human varieties, as well as their average weight and height, were important components of their race studies.

Natural history classificatory practices were also an important part of early ethnological and anthropological observations. The ability to distinguish general and subtle physical differences between the races, as well as their cultural peculiarities and similarities, was an integral component of nineteenth-century methodologies. Books such as Prichard's *Researches* and Latham's *Varieties of Man* divided different ethnicities into groups based on their physical structure, geographic location and language. These taxonomical methods were appropriated from the works of naturalists such as Linnaeus, Buffon and Blumenbach. We also saw how figures such as Hunt and Huxley strategically used the Baconian method of induction in their studies to substantiate their research claims. In both cases, illustrative tables were an important aspect of their didactic literature.

However, observational practices involve a three-part process. By breaking down the way in which early practitioners engaged with their object of study, we can see how their training and interests in natural history and medicine influenced their observational practices as a whole. It also shows the sophistication of these methods and demonstrates that nineteenth-century researchers were not simply anthologizing older material but were producing novel scientific research. Under this model, the first step for researchers is to collect their data either directly, as was the case with figures such as Knox and Darwin, or indirectly through the use of informants, as was the case with Prichard and Hodgkin. This data not only includes written reports such as travel accounts by surgeons and naturalists, but also objects such as cultural artefacts and skulls. Secondly, researchers make sense of their data by using various analytical techniques, from comparative anatomy and natural history classificatory practices used by figures such as Lawrence and Hunt, to developmental theories such as Darwinian evolution and Tylorian cultural developmentalism. This process of interpreting ethnographic data can also include the use of tools and instruments such as cameras, as we saw with schemes such as Huxley's 'well-considered-plan' in the late 1860s. Thirdly, researchers represent their results through various forms of media. These include lectures and handbooks for specialized and non-specialized audiences, imagery such as paintings, engravings and photographs, and museum displays, from small nineteenth-century penny shows in London to large exhibits such as the Crystal Palace during the 1850s.

Throughout this book, I have been preoccupied with the observational practices of Victorian anthropology. However, historians can gain a richer understanding of any discipline's history by applying my analytical framework to their studies. Broadly speaking, observation was a central component of all scientific disciplines in the nineteenth century. Take for example the techniques utilized by Victorian anatomists. Historians of medicine have begun to look much more closely at the history of medical education. Andrew Cunningham shows us in his book *The Anatomist Anatomis'd* (2010) how late eighteenth-century medical students learned about their subject in schools. He provides us with a rich overview of the medical world in this period. Similarly, A. W. Bates has shown us in his work *The Anatomy of Robert Knox* (2010) the importance of Knox's surgical lectures for medical students training in Edinburgh.[25] In her book *The Making of Mr. Gray's Anatomy* (2008), Ruth Richardson examines how Henry Gray's anatomy textbook transformed nineteenth-century medical instruction.[26] Building upon and moving beyond these studies, it would be further instructive for historians of medicine – who are interested in the history of anatomy – to incorporate my framework for analysing observational practices into their work.

By using my three-part model to examine how nineteenth-century physicians and surgeons went about observing, analysing and representing the human body, historians would have a much more comprehensive understanding of historical medical practices. Similar interpretive frameworks for analysing observational practices have already yielded impressive results in the works of historians such as Anne Secord, who has focused on botany, and Lorraine Daston and Peter Galison, who have examined the natural sciences.[27]

## *Disciplinary Development*

Another central theme in this book was disciplinary development. Each of the figures discussed attempted to refine and transform the practices of the discipline. It was their hope to place the study of human variation on an equal footing with other sciences. In Chapter 1 I connected the emergence of ethnology, as a specialized field of enquiry, to the publication of Prichard's book the *Researches* and Lawrence's lectures on the *Natural History of Man*. Both of these texts represented early attempts by nineteenth-century practitioners to establish a discipline strictly devoted to the study of human diversity and separate from other fields of study, which shared similar research interests such as geography and orientalism. Thus Prichard and Lawrence outlined for their audiences the intellectual scope and aims of ethnology.

Disciplinary development was also an important theme in Chapter 2, where I focused on the founding of the ESL. Their society meetings provided early researchers with a place to refine and negotiate their enquiries. It demonstrated to other disciplines that ethnologists wanted to raise the scientific standards of the discipline by providing its practitioners with an opportunity to present and discuss their work with fellow scholars interested in human variation topics, and by providing them with a specialized journal in which ethnologists could publish their results, thus further expanding their audiences. The establishment of the ESL was also important for the development of the discipline because the society played a key role in building informant networks around the world. These informants were significant contributors to ethnological (and later anthropological) studies because they provided researchers living in Britain with ethnographic data to support their theories. Prichard, Latham and Hunt never left the shores of Europe. As a result, all of their ethnological and anthropological enquiries of extra-Europeans used the reports of informants. However, not all nineteenth-century ethnology and anthropology focused on human populations living beyond the boundary of Europe. The ethnography of Europe was a significant component of Victorian race studies. Both Prichard and Latham wrote important works on European diversity. There were other instances where Victorian practitioners conducted locally based studies that observed different ethnicities first-hand. Knox visited Jewish neighbourhoods in London and Amsterdam to collect his ethnographic data.[28]

Another instance where disciplinary development was an important part of my analysis was in Chapter 4, where I revisited the anthropological schism of the 1860s. There are several key points that emerged in this case study which relate to this theme. For both Hunt and Huxley, it was important that anthropology and ethnology clearly defined their research scopes and aims. Hunt and Huxley believed they were strengthening anthropology's and ethnology's claims to specialized types of knowledge. We also saw in Chapter 4 that Hunt played an important part in expanding anthropological literature in Britain. During the short existence of the ASL, he produced two anthropological journals and published translated editions of anthropological texts from France and Germany for British practitioners to incorporate into their researches. At the same time, we also saw how Huxley's 'well-considered-plan' helped to transform the discipline by further expanding its informant networks. This in turn brought a new kind of ethnographic imagery back to the metropole for British researchers to use in their enquiries.

As David Kaiser has shown us in his article 'Training and the Generalist's Vision in the History of Science' (2005), a major research area within the history of science is disciplinary development.[29] By focusing on this theme, it enabled me to connect my story about observational practices in anthropology to larger historiographic issues outside of my subfield. This is important because so much of the history of anthropology lies on the periphery of various historical subfields – including history of science, cultural history and imperial history – as well as anthropology. Bringing disciplinary development to the fore of my book has also allowed me to show how my work builds upon and moves beyond important historiographic literature, such as Jack Morrell and Arnold Thackray's polemic book *Gentlemen of Science* (1981) and more recent works such as Graeme Gooday's *The Morals of Measurement* (2004) and Ralph O'Connor's *The Earth on Show* (2007).[30] Each of these texts has examined the way in which practitioners from different research fields have attempted to transform the scientific standards of their disciplines.

## *Location and Research Practices*

The third major theme I want to discuss is the significance of location for anthropology. From the onset of this book, I argued that the location where practitioners observe their object of study is an important part of their observational practices. However, it is not the defining feature that makes a work anthropological. It is equally important for anthropologists to do preparatory readings when beginning new projects. Anthropologists can also gain a substantial amount by reading the second-hand reports of others who have visited a region and its people. Photographs and films can further provide useful insights into different cultures. Thus fieldwork, in combination with other tools and methodologies, offers anthropologists an observational arsenal that allows them to engage with their object of study. Nineteenth-century practitioners who were interested in human diversity were no different, and they incorporated a multitude of different techniques and apparatuses into their ethnological and anthropological studies. Moreover, it is not necessary for the research location to be in an exotic land. Within the context of the Victorian era, there are numerous examples of practitioners conducting ethnographic studies within Europe. Latham, for instance, spent time in rural Norway collecting philological evidence for his ethnological studies.[31]

The theme of location also featured in my five case studies in other ways. For instance, all of the figures discussed in this book resided in Britain, with most working in London at some point in their careers. The majority of early ethnologists were unable to travel abroad and collect their own ethnographic data because of commitments back in Britain. Thus they were reliant on the assistance of informants living around the world to furnish them with data to incorporate into their studies of extra-Europeans. Because London was at the centre of Britain's empire, it was an ideal place for ethnologists (and later anthropologists) to reside. It provided them with easier access to the latest ethnographic material returning to the metropole, as well as many opportunities for gainful employment. During the first half of the nineteenth century, it was virtually impossible for researchers interested in human diversity to make a living as full-time ethnologists or anthropologists. Most of my figures (with the exception of Tylor) had careers in other fields. Prichard, Latham and Hodgkin were physicians, and Lawrence, Knox and Huxley were prolific lecturers – among other things.

I also discussed issues relating to location in Chapter 3, when I examined the importance of *ex situ* observational sites. For this analysis, I looked at Latham's directorship of the ethnological department at the Crystal Palace, as well as Knox's visits to Jewish neighbourhoods in London and Amsterdam and his trips to the English countryside where he studied Gypsy communities. In both cases, these two figures were attempting to strengthen the discipline's research techniques by establishing important sites within the United Kingdom for practitioners to further engage with their object of study. As demonstrated, the ethnological exhibit at the Crystal Palace was an important locally based training site for anyone interested in contributing to the sciences of Man. Conversely, Knox's direct engagement with Jews and Gypsies is another example of how early practitioners attempted to increase their first-hand knowledge of different ethnicities by doing a type of field study *avant la lettre* inside of Europe.

Throughout this book, I also provided multiple examples of early ethnologists and anthropologists travelling abroad and collecting their own ethnographic data *in situ*. For example, I showed how Tylor spent several months travelling through Mexico and studying the indigenous people and their history. Darwin also collected ethnographic data during the *Beagle* voyage, and his descriptions of his encounters with groups such as the Fuegians were important evidentiary sources for practitioners living in Britain. Knox also possessed some first-hand experience of observing indigenous populations in their own environment from his time stationed in South Africa during the Cape Frontiers War. This experience shaped his observational framework, leading him to argue that all ethnological studies should be based primarily on directly observable evidence.

Recently, the theme of location has been an important component of works by historians of empire, as well as by visual anthropologists. The significance of the metropole in anthropological knowledge circuits builds upon – and moves beyond – the work of Catherine Hall in her book *Civilising Subjects* (2002) and Zoë Laidlaw in her work *Colonial Connections 1815–1845* (2005).[32] James Secord has also discussed the ways in which knowledge networks function in his article 'Knowledge in Transit' (2004). Secord has provided us with an important explanatory framework for how ideas transfer across different boundaries, both physical and conceptual.[33] My focus on the relationship between location and observational practices within Victorian anthropology further connects the disciplinary history of the sciences of Man to a larger body of secondary literature on the global history of science.

## *Visualizing Culture*

Vision is another significant theme throughout this book. In particular, I examined in detail the history of visual anthropology in the first half of the nineteenth century, by looking at the ways in which early researchers interested in human diversity represented different races through various forms of media. This included lithographs such as the images discussed in my examination of Latham's *Varieties of Man*, museum displays such as the ethnological department at the Crystal Palace, and photographs such as those collected by colonial officials through Huxley's 'well-considered-plan' in the 1860s. My focus on observational practices also connects this book to recent literature in visual anthropology by adding a historical component to these discussions. For example, Cristina Grasseni's edited volume *Skilled Visions* (2007) has several essays on 'visual enskilment' and location, although they predominantly look at contemporary anthropological research activities.[34] Anna Grimshaw has also looked at 'ways of seeing' in her book *The Ethnographer's Eye* (2001). However, her study focuses on twentieth-century practices.[35] My focus on the history of observational practices in anthropology also connects my research to the work of Lorraine Daston and Elizabeth Lunbeck and their book *Histories of Scientific Observation* (2011).[36] To sum up, by looking at these four major themes, this book opens up new research questions for scholars from a broad range of specializations, including history of anthropology, history of race, history medicine, imperial history, cultural anthropology and visual anthropology. Moreover, by integrating my analytical framework – with its focus on observational practices – into other historical and cultural studies, new light can be shed on a variety of topics.

## Beyond the Armchair: New Directions in the History of Anthropology

Because this book focused primarily on researchers in London, it would be instructive to consider in more detail the informants who collected the data for these metropolitan researchers. Such an approach would build upon recent scholarship that seeks to shift traditional historiographical focuses away from Europe and towards the colonial world. For example, in Sujit Sivasundaram's book *Nature and the Godly Empire* (2005), he shows us the important role missionaries played in disseminating science throughout the world.[37] In terms of observational practices, missionary reports would be equally valuable in showing another group of travellers who contributed to anthropological research in the nineteenth century. We can also learn more about data collection by looking at the correspondences of key figures such as Huxley and Darwin, and analysing the responses they received to research queries from their correspondents living abroad. Because we have such a rich archive for both of these figures, we are fortunate that many of their informants' responses survive. There are other underexplored resources that could further enrich such an investigation. Governor reports from the colonies would be equally valuable in showing another group of travellers who contributed to transformations in anthropology during the nineteenth century.

Building on the ground-breaking work of Sadiah Qureshi, another research area that could be expanded in relation to anthropological observational practice is the way early researchers taught non-specialist audiences about human diversity.[38] In this book, I discussed the significance of ethnographic displays – such as the ethnological department at the Crystal Palace – as important sites for training practitioners how to examine human varieties in a specialized way. However, it would be useful to see how this information was communicated to other parts of Victorian society. Institutions such as the British Museum and the Egyptian Hall were important sites of learning. Equally significant, however, were the cheap penny shows that lined the streets of London. Looking at how human diversity was taught to popular audiences at these sites would enrich our understanding of anthropological knowledge and practice in the nineteenth century. In particular it would give us a better understanding of how non-specialists observed, analysed and represented human diversity. Building on the work of Jonathan Topham and James Secord, it is also important to see how these shows – and anthropological literature generally – were discussed and reviewed in the periodical press.[39] My small case study that analysed the reviews of Knox's and Latham's works yielded interesting results. It showed, to a certain extent, the reception of these books. Broadening such an examination by incorporating even more nineteenth-century titles would greatly expand our knowledge of anthropology's history.

Another theme that could be further developed is a comparative study of different anthropological traditions. Within this book, I examined the influence of French anthropology on the British tradition. It would be instructive to examine in more detail the observational practices of nineteenth-century French practitioners and compare their methods with those of the British. One characteristic that particularly stands out is how French and British researchers organized their division of labour. French ethnology and anthropology often involved a single individual responsible for both collecting data *in situ* and writing accounts based on an analysis of the evidence available. The British tradition, by contrast, relied more heavily on networks of informants who collected data abroad and sent this material back to Britain for researchers to analyse and turn into reports. The armchair paradigm is not a major theme within the French anthropological tradition.[40] Perhaps this is because the French did not have as clear a division between collectors and analysers as was the case with their British counterparts. Moreover, the French were engaging much earlier in an identifiable form of ethnographic fieldwork.[41]

A comprehensive study of observational practices within the North American context would also further enrich our understanding of early ethnological and anthropological methodologies. Unlike their European counterparts, American anthropologists had direct access to Native American tribes and African slaves. Thus American scholars were able to incorporate into their writings more first-hand experience of their interactions with extra-Europeans. The work of figures such as Josiah Nott, who visited plantations in the Southern United States in order to measure the skulls of slaves, demonstrates the lengths to which nineteenth-century American anthropologists attempted to engage directly with their object of study. Because nineteenth-century American anthropologists had the opportunity to interact daily with extra-European peoples, it is difficult to characterize their methodologies as simply armchair ruminations.

Finally, there is still much work to be done within the British context. There are several key figures not discussed in the preceding chapters, consideration of whom would greatly expand our understanding of early observational practices within the sciences of Man. For example, the colonial officer and writer Richard Francis Burton (1821–90) had a comprehensive knowledge of several Asian cultures from his travels through the East. His pilgrimage to Mecca disguised as an Afghani Pashtun can be interpreted as a form of participant observation *avant la lettre*. His immersion in Middle Eastern customs and rituals allowed him to see aspects of the world through the eyes of extra-European peoples. The Victorian polymath Francis Galton (1822–1911) would be equally illuminating for a study on anthropological observational practices. Galton not only possessed primary experience with indigenous people from his travels through Africa and the Middle East during the mid-1840s and early 1850s, but he also organized a scheme in 1877 to photograph different groups of people for anthropometric studies.[42] It built upon the earlier work of figures such as Huxley.

On a final note, I would like to return to the 'presentism vs. historicism' debate. The history of anthropology has long resided on the fringe of two disciplines. On the one hand, its focus on the historical practices of anthropologists has at times attracted the attention of social and cultural historians; on the other hand, its emphasis on demonstrating its applicability to contemporary anthropologists has made the subfield popular among social scientists. However, as Stocking, Kuklick and others have routinely argued in their work, it is the interdisciplinary nature of the history of anthropology that gives the subfield its unique position and pushes its scholars to ask new questions.[43] Building upon this tradition, it was my hope that a focus on observational practices would appeal to researchers from both disciplines.

In this book, I looked at the whole observational activity conducted during the first half of the nineteenth century, emphasizing its commonalities with later twentieth-century methods. By showing these continuities, I demonstrated how knowledge of the discipline's past has a bearing on contemporary anthropological practices. What it means to be an anthropologist encompasses a broad range of activities, including reading, writing, collecting, analysing and participating through fieldwork. It was my hope that by combining 'historicist' and 'presentist' interests, my book would shed new light on the subfield of history of anthropology and help build new areas of research. The disciplinary shift that occurred at the beginning of the twentieth century, which saw anthropology reposition itself as a field-based science, was part of a longer process that aimed to transform the way its researchers engaged with their object of study. The introduction of fieldwork as the cornerstone of the discipline in the 1910s and after thus emerges as one more tool added to the arsenal of the skilled observer.

# NOTES

## Introduction

1. E. B. Tylor, *Anthropology: An Introduction to the Study of Man and Civilisation* (London: Macmillan, 1881), pp. 336–7.
2. B. Malinowski, *Argonauts of the Western Pacific: An Account of the Native Enterprise and Adventure in the Archipelagoes of Melanesian New Guinea*, 2nd edn (London: George Routledge & Sons, Ltd, 1932), p. 2.
3. Ibid., pp. vii–viii.
4. A. Gupta, and J. Ferguson, 'Discipline and Practice: "The Field" as Site, Method, and Location in Anthropology', in A. Gupta and J. Ferguson (eds), *Anthropological Locations: Boundaries and Grounds of a Field Science* (Berkeley, CA: University of California Press, 1997), pp. 1–46, on p. 1.
5. M. Angrisino and J. Crane, *Field Projects in Anthropology: A Student Handbook* (Morristown, NJ: General Learning Press, 1974), p. v
6. J. Urry, 'A History of Field Methods', in R. F. Ellen (ed.), *Ethnographic Research: A Guide to General Conduct* (London: Academic Press, 1984), pp. 27–61, on p. 27.
7. H. Kuklick, 'After Ishmael: The Fieldwork Tradition and its Future', in A. Gupta and J. Ferguson (eds), *Anthropological Locations: Boundaries and Grounds of a Field Site* (Berkeley, CA: University of California Press, 1997), pp. 47–65, on p. 47.
8. Ibid., p. 49.
9. Gupta and Ferguson, 'Discipline and Practice', p. 1.
10. B. Tedlock, 'From Participant Observation to the Observation of Participation: The Emergence of Narrative Ethnography', *Journal of Anthropological Research*, 41 (1991), 61–94, on p. 69.
11. Ibid.
12. J. Urry, 'Notes and Queries on Anthropology and the Development of Field Methods in British Anthropology, 1870–1920', *Proceedings of the Royal Anthropological Institute of Great Britain and Ireland*, 1972 (1972), pp. 45–57, on p. 45.
13. H. Kuklick, 'Personal Equations: Reflections on the History of Fieldwork, with Special Reference to Sociocultural Anthropology', *Isis*, 102 (2011), pp. 1–33.
14. Kuklick, 'After Ishmael', p. 51.
15. Ibid., p. 49.
16. For more information on anthropology and its links to colonial development in the early twentieth century, consult: H. Kuklick, 'The Sins of the Fathers: British Anthropology and African Colonial Administration', *Research in Sociology of Knowledge, Science and Art*, 1 (1978), pp. 93–119; H. Kuklick, *The Savage Within: The Social History of Brit-*

*ish Anthropology, 1885–1945* (Cambridge: Cambridge University Press, 1991), pp. 41–3; and H. Kuklick, 'The British Tradition', in H. Kuklick (ed.), *A New History of Anthropology* (Oxford: Blackwell Publishing, 2008), pp. 52–78, on pp. 60–1. For more on Malinowski's career, see A. Kuper, *Anthropology Anthropologists: The Modern British School* (London: Routledge, 1993), pp. 1–35.

17. F. Erikson, 'What Makes School Ethnography "Ethnographic"?', *Anthropology & Education Quarterly*, 14 (1984), pp. 51–66, on p. 53.
18. Tedlock, 'From Participant Observation', pp. 69–70.
19. A. Grimshaw, *The Ethnographer's Eye: Ways of Seeing in Modern Anthropology* (Cambridge: Cambridge University Press, 2001), p. 4.
20. R. F. Ellen, 'Introduction', in R. F. Ellen (ed.), *Ethnographic Research: A Guide to General Conduct* (London: Academic Press, 1984), pp. 1–12, on p. 10.
21. Ibid., p. 3.
22. Ibid., p. 10.
23. Grimshaw, *The Ethnographer's Eye*, p. 6.
24. F. Barth, 'Britain and the Commonwealth', in F. Barth, A. Gingrich, R. Parkin and S. Silverman, *One Discipline Four Ways: British, German, French and American Anthropology* (Chicago, IL: University of Chicago Press, 2005), pp. 3–57, on. p. 6.
25. Ibid., p. 9.
26. Ibid., p. 4.
27. Kuklick, *The Savage Within*, p. 6.
28. Ibid.
29. Kuklick, 'After Ishmael', p. 53. For other examples within the historiography, see Barth, 'Britain and the Commonwealth', pp. 9–10; and Gupta and Ferguson, 'Discipline and Practice', p. 7.
30. Kuklick, *The Savage Within*, pp. 14–16.
31. R. Kohler, *Landscapes and Labscapes: Exploring the Lab-Field Border in Biology* (Chicago, IL: University of Chicago Press, 2002), p. 2.
32. Urry, 'A History of Field Methods', p. 27.
33. Ibid., pp. 37–8.
34. For more information on the politics of writing, see: J. Clifford and G. E. Marcus (eds), *Writing Culture: The Poetics and Politics of Ethnography* (Berkeley, CA: University of California Press, 1986); H. White, *Tropics of Discourse: Essays in Cultural Criticism* (Baltimore, MD: Johns Hopkins University Press, 1978); H. White, *The Content of the Form: Narrative Discourse and Historical Representation* (Baltimore, MD: Johns Hopkins University Press, 1987); A. Krupat, *Ethnocentrism: Ethnography, History, Literature* (Berkeley, CA: University of California Press, 1992); and A. James et al., *After Writing Culture: Epistemology and Praxis in Contemporary Anthropology* (London: Routledge, 1997).
35. G. Stocking, 'On the Limits of "Presentism" and "Historicism" in the Historiography of the Behavioural Sciences', in G. Stocking, *Race, Culture, and Evolution: Essays in the History of Anthropology* (London: Blackwell, 1982), pp. 1–12, on p. 3.
36. Ibid., p. 6.
37. G. Stocking, *Victorian Anthropology* (New York: Free Press, 1987), p. 285.
38. H. Kuklick, 'Introduction', in H. Kuklick (ed.), *A New History of Anthropology* (Oxford: Blackwell Publishing, 2008), pp. 1–16, on p. 1.
39. Ellen, 'Introduction', p. 1.
40. Ibid., p. 3.
41. Gupta and Ferguson, 'Discipline and Practice', p. 2.

42. Ibid.
43. Ibid.
44. S. Schaffer, *From Physics to Anthropology and Back Again* (Cambridge: Prickly Pear Press, 1994), pp. 6–7.
45. For example, Tedlock, 'From Participant Observation', p. 69. Haddon also divided the history of anthropology into three methodological epochs; see A. C. Haddon and A. H. Quiggin, 'Introduction', in A. C. Haddon and A. H. Quiggin, *History of Anthropology* (London: G. P. Putman's Sons, 1910), pp. ix–xiv, on p. x.
46. Schaffer, *From Physics to Anthropology and Back Again.* For more on Haddon and anthropology, see: A. A. Rodan, 'Looking at Anthropology from a Biological Point of View: Haddon's Metaphors on Anthropology', *History of the Human Sciences*, 5 (1992), pp. 21–32; J. Urry, 'Englishmen, Celts and Iberians: The Ethnographic Survey of the United Kingdom', in G Stocking (ed.), *Functionalism Historicised: Essays in British Anthropology* (Madison, WI: University of Wisconsin Press, 1984), pp. 83–105; and P. Gathercole, 'Cambridge and the Torres Straits, 1888–1920', *Cambridge Anthropology*, 3 (1977), pp. 22–31.
47. For examples of British instructional guides from the first half of the nineteenth century, see: H. Martineau, *How to Observe: Morals and Manners* (London: Charles Knight and Co., 1838); J. C. Prichard, 'Ethnology', in J. Herschel (ed.), *A Manual of Scientific Enquiry: Prepared for the Use of Officers in Her Majesty's Navy and Travellers in General* (London: John Murray, 1849), pp. 423–40; T. Hodgkin and R. Cull, 'A Manual of Ethnological Inquiry', *Journal of the Ethnological Society of London*, 3 (1854), pp. 193–208.
48. D. Bleichmar, 'Training the Naturalist's Eye in the Eighteenth Century: Perfect Global Visions and Local Blind Spots', in C. Grasseni (ed.), *Skilled Visions: Between Apprenticeship and Standards* (Oxford: Berghahn, 2007), pp. 166–90, on p. 168.
49. For a comprehensive study of observational practices in the nineteenth century, see ch. 3 of A. Secord, *Artisan Naturalists: Science as Popular Culture in Nineteenth-Century England* (PhD thesis, University of London, 2002), pp. 135–206.
50. L. Daston and P. Galison, *Objectivity* (London: Zone Books, 2007), p. 19.
51. J. Urry, *Before Social Anthropology: Essays on the History of British Anthropology* (Chur: Harewood Academic Publishers, 1993), p. 2. For more information on observational practices within modern anthropology, consult C. Grasseni (ed.), *Skilled Visions: Between Apprenticeship and Standards* (Oxford: Berghahn, 2007); and Grimshaw, *The Ethnographer's Eye.*
52. The anatomy of races was a central theme in the works of many early ethnologists and anthropologists. For example, see J. C. Prichard, *Researches into the Physical History of Man* (John and Arthur Arch, 1813) (unless otherwise noted, all subsequent citations are to this edition); and W. Lawrence, *Lectures on Physiology, Zoology, and the Natural History of Man, Delivered at the Royal College of Surgeons* (Salem, NY: Foote and Brown, 1828).
53. Daston and Galison, *Objectivity*, p. 23.
54. For more, see S. Schaffer and J. Delbourgo et al., *The Brokered World: Go-Betweens and Global Intelligence 1770–1820* (Sagamore Beach, MA: Science History Publications, 2009).
55. S. Shapin, *A Social History of Truth: Civility and Science in Seventeenth-Century England* (Chicago, IL: University of Chicago Press, 1994), p. xxv.
56. Ibid.
57. For more information on the reliability of testimony in substantiating knowledge claims, consult C. A. J. Coady, *Testimony: A Philosophical Study* (Oxford: Oxford University Press, 1992).

58. Secord, *Artisan Naturalists*, p. 13.
59. Ibid., p. 14. See also E. P. Thompson, 'Folklore, Anthropology and Social History', *Indian Historical Review*, 3 (1977), pp. 247–66, on p. 248.
60. G. Stocking, 'Retrospective Prescriptive Reflections', in G. Stocking, *The Ethnographer's Magic and Other Essays in the History of Anthropology* (Madison, WI: University of Wisconsin Press, 1992), pp. 3–11, on pp. 3–4.
61. Prichard, *Researches*; and Lawrence, *Natural History of Man*.
62. G. Long (ed.), *Penny Cyclopaedia of the Society for the Diffusion of Useful Knowledge*, 12 vols (London: Charles Knight, 1833–43), vol. 2, p. 97.
63. For more discussion on the debates surrounding the disciplinary label of anthropology in the first half of the nineteenth century, see G. Stocking, 'What's in a Name? The Origins of the Royal Anthropological Institute (1837–71)', *Man*, 6 (1971), pp. 369–90.
64. T. Barfield (ed.), *The Dictionary of Anthropology* (Oxford: Blackwell Publishing, 1997), p. 157.
65. J. C. Prichard, *The Natural History of Man: Comprising Inquiries into the Modifying Influence of Physical and Moral Agencies on the Different Tribes of the Human Family*, 1st edn (London: H. Baillière, 1843), p. 132. See also Barfield, *Dictionary of Anthropology*, p. 157.

## 1 Founding the Sciences of Man: The Observational Practices of James Cowles Prichard and William Lawrence

1. Lawrence, *Natural History of Man*, p. 213.
2. For historiographic accounts of Prichard, see: G. Stocking, 'What's in a Name?'; G. Stocking, 'From Chronology to Ethnology: James Cowles Prichard and British Anthropology 1800–1850', in J. C. Prichard, *Researches into the Physical History of Man*, ed. G. Stocking (Chicago, IL: University of Chicago Press, 1973), pp. ix–cx; J. Efron, *Defenders of Race: Jewish Doctors and Race Science in Fin-de-Siècle Europe* (New Haven, CT: Yale University Press, 1994), pp. 44–5; H. F. Augstein, *James Cowles Prichard's Anthropology: Remaking the Science of Man in Early Nineteenth-Century Britain* (Amsterdam: Rodopi B.V., 1999); R. Kenny, 'From the Curse of Ham to the Curse of Nature: The Influence of Natural Selection on the Debate on Human Unity before the Publication of the "Descent of Man"', *British Journal for the History of Science*, 40 (2007), pp. 367–88. For accounts of Lawrence, see: J. Goodfield-Toulmin, 'Some Aspects of English Physiology: 1780–1840', *Journal of the History of Biology*, 2 (1969), pp. 283–320; L. S. Jacyna, 'Immanence or Transcendence: Theories of Life and Organisation in Britain, 1790–1835', *Isis*, 74 (1983), pp. 311–29; Stocking, *Victorian Anthropology*, pp. 43, 66, 97; A. Desmond, *The Politics of Evolution: Morphology, Medicine, and Reform in Radical London* (Chicago, IL: University of Chicago Press, 1989), pp. 117–21; and Efron, *Defenders of Race*, pp. 44–5.
3. W. B. Carpenter, 'Art V.-1 Researches into the Physical History of Mankind', *Edinburgh Review*, 88 (1848), 429–87, on pp. 486–7.
4. Ibid., p. 487.
5. Anon., 'Review – Lectures on Physiology, &c.', *Imperial Magazine*, 4 (1822), pp. 668–74, on p. 668.
6. Lawrence's *Natural History of Man* was first published in 1819. However, because of its controversial material it had a limited publishing run. Nine unofficial reprinted editions were produced between 1819 and 1848. For more, see Desmond, *The Politics of Evolution*, pp. 117–21.

7. Anon., 'Researches into the Physical History of Mankind', *Athenaeum*, 776 (1842), pp. 797–98, on p. 797.
8. Both Augstein and Stocking discuss in detail the influence of Christianity on Prichard's monogenetic paradigm. For more, see ch. 3 of Augstein, *James Cowles Prichard's Anthropology*, pp. 57–128, and Stocking, 'From Chronology to Ethnology'.
9. Augstein, *James Cowles Prichard's Anthropology*, p. 2.
10. T. Hodgkin, 'Obituary of Dr. Prichard', *Journal of the Ethnological Society of London*, 2 (1850), pp. 182–207, on p. 184; and H. F. Augstein, 'Prichard, James Cowles (1786–1848), Physician and Ethnologist', *ODNB*.
11. For more information on Quakers at Cambridge and Oxford, see G. Cantor, *Quakers, Jews and Science: Religious Responses to Modernity and the Sciences in Britain 1650–1900* (Oxford: Oxford University Press, 2005), pp. 82–101.
12. For information on the history of medicine at the University of Edinburgh, see: Augstein, *James Cowles Prichard's Anthropology*, pp. 3–5; L. S. Jacyna, *Philosophical Whigs: Medicine, Science and Citizenship in Edinburgh, 1789–1848* (London: Routledge, 1994); and L. Rosner, *Medical Education in the Age of Improvement: Edinburgh Students and Apprentices, 1760–1826* (Edinburgh: Edinburgh University Press, 1991). For information on Quakers at Edinburgh, see Cantor, *Quakers, Jews and Science*, pp. 63–71.
13. L. Rosner, 'Monro, Alexander, Secundus, (1733–1817), Anatomist', *ODNB*.
14. For more on the medical programme and extramural schools in Edinburgh, see: Augstein, *James Cowles Prichard's Anthropology*, pp. 3–5; Jacyna, *Philosophical Whigs*, pp. 85–92; Rosner, *Medical Education in the Age of Improvement*; A. Desmond and J. Moore, *Darwin* (London: Michael Joseph, 1991), pp. 21–31; and A. Desmond and J. Moore, *Darwin's Sacred Cause: Race, Slavery and the Quest for Human Origins* (London: Penguin Books, 2009), pp. 27–48.
15. For examples of Prichard's use of Blumenbach, see Prichard, *Researches*, pp. 46, 55, 68, 81, 241 and 271. There is a large body of work on Prichard and the influence of continental natural history and philosophy on his ethnological paradigm. The two best sources are: Stocking, 'From Chronology to Ethnology'; and Augstein, *James Cowles Prichard's Anthropology*.
16. J. C. Prichard, 'Of the Varieties of the Human Race', *Records of the Royal Medical Society of Edinburgh*, 58 (1807–8), pp. 87–134; and Augstein, *James Cowles Prichard's Anthropology*, p. 4.
17. Stocking, 'From Chronology to Ethnology', pp. xv–xviii; and Augstein, *James Cowles Prichard's Anthropology*, p. 2.
18. Augstein, *James Cowles Prichard's Anthropology*, p. 2.
19. J. C. Prichard, *A Treatise on Insanity* (London: Marchant, 1833); J. C. Prichard, *A Treatise on Insanity, and Other Disorders Affecting the Mind* (London: Sherwood, 1835); and J. C. Prichard, *On the Different Forms of Insanity, in Relation to Jurisprudence. Designed for the Use of Persons Concerned in Legal Questions Regarding Unsoundness of Mind* (London: H. Baillière, 1842). For a complete bibliographic list of all of Prichard's writings on mental illness and ethnology, see G. Stocking, 'Bibliography of James Cowles Prichard', in J. C. Prichard, *Researches into the Physical History of Man*, ed. G. Stocking (Chicago, IL: University of Chicago Press, 1973), pp. cxi–cxviii. For information on the sources of Prichard's primary materials, see G. Stocking, 'Guide to Prichard's References', in J. C. Prichard, *Researches into the Physical History of Man*, ed. G. Stocking (Chicago, IL: University of Chicago Press, 1973), pp. cxix–cxliv.

20. For more, see: Goodfield-Toulmin, 'Some Aspects of English Physiology'; Jacyna, 'Immanence or Transcendence'; Stocking, *Victorian Anthropology*, pp. 43, 66, 97; Desmond, *The Politics of Evolution*, pp. 117–21; and Efron, *Defenders of Race*, pp. 44–5.
21. L. S. Jacyna, 'Lawrence, Sir. William, First Baronet (1783–1867), Surgeon', *ODNB*.
22. J. F. Blumenbach, *A Short System of Comparative Anatomy*, trans. W. Lawrence (London: Longman and Co., 1807).
23. Lawrence, *Natural History of Man*, pp. 112–13.
24. Ibid., pp. iii–iv.
25. W. Lawrence, *An Introduction to Comparative Anatomy and Physiology; Being the Two Introductory Lectures Delivered at the Royal College of Surgeons, on the 21st and 25th of March, 1816* (London: J. Callow Medical Booksellers, 1816), p. 166.
26. Desmond, *The Politics of Evolution*, p. 117. For more information on Lawrence and materialism, see: Goodfield-Toulmin, 'Some Aspects of English Physiology'; Jacyna, 'Immanence or Transcendence'; and Desmond, *The Politics of Evolution*, pp. 117–21, 255–7.
27. T. H. Huxley, 'On the Methods and Results of Ethnology', *Fortnightly Review*, 1 (1865), pp. 257–77, on pp. 263–5; and C. Darwin, 'Transmutation of Species (1838.02–1838.07)', Notebook C, Charles Darwin Archive, University Library, University of Cambridge (hereafter Notebook C), p. 204.
28. C. Linnaeus, *The Elements of Botany: Containing the History of Science*, trans. H. Rose (London: Cadell, 1775), p. 269.
29. Prichard, *Researches*, p. 1.
30. Ibid., pp. 1–2.
31. Ibid., p. 2.
32. Ibid., p. 3.
33. Ibid., pp. 2–3.
34. Ibid., p. 6.
35. Ibid., p. 7.
36. Ibid.
37. Ibid.
38. Ibid., pp. 9–10.
39. Lawrence, *Natural History of Man*, p. 106.
40. Ibid., p. 107.
41. Lawrence did not identify any specific scholar in his critiques of human variation studies.
42. Lawrence, *Natural History of Man*, p. 107.
43. Ibid., p. 118.
44. Ibid., p. 213.
45. Ibid.
46. Ibid., pp. 215–16.
47. Ibid., p. 226.
48. Prichard, *Researches*, pp. 9–10.
49. Lawrence, *Natural History of Man*, p. 226.
50. Ibid., pp. 227, 233.
51. Ibid., p. 228.
52. Ibid., p. 234.
53. Ibid., p. 235.
54. Ibid., p. 112–13.
55. Prichard, *Researches*, pp. 17–18.
56. Ibid., p. 20.

57. Ibid.
58. Ibid., p. 21.
59. Ibid., p. 21–2.
60. Ibid., p. 22–3.
61. Ibid., p. 23.
62. Ibid.
63. Ibid., p. 24.
64. Ibid., pp. 24–5.
65. For more information on Prichard's theory of hereditary transmission, see ibid., pp. 25–32.
66. Ibid., p. 27.
67. Ibid., p. 33.
68. Ibid., pp. 34–5.
69. For more, see Augstein, *James Cowles Prichard's Anthropology*, p. 61.
70. Prichard, *Researches*, p. 46.
71. Ibid., pp. 55–6.
72. Ibid., p. 56.
73. Ibid.
74. See Prichard, *Researches*, pp. 56–7; and J. F. Blumenbach, *The Anthropological Treatises of Johann Friedrich Blumenbach, Late Professor at Göttingen and Court Physician to the King of Great Britain*, trans. T. Bendyshe (London: Longman, Green, Longman, Roberts, & Green, 1865), p. 237.
75. Prichard, *Researches*, p. 57; and Blumenbach, *Anthropological Treatises*, p. 238.
76. Prichard, *Researches*, p. 57; and Blumenbach, *Anthropological Treatises*, p. 238.
77. Prichard, *Researches*, pp. 57–8.
78. For Blumenbach's description of American and Malay skulls, see Blumenbach, *Anthropological Treatises*, p. 238.
79. Lawrence, *Natural History of Man*, p. 236.
80. Ibid., p. 242.
81. Ibid., p. 243.
82. Ibid., p. 245.
83. Ibid.
84. Ibid., p. 251.
85. Ibid.
86. Ibid.
87. Ibid.
88. Ibid.
89. Ibid.
90. Ibid., p. 252.
91. Ibid.
92. Prichard, *Researches*, pp. 25–32.
93. Lawrence, *Natural History of Man*, p. 254.
94. Ibid., pp. 268–9.
95. For more on Prichard's hair taxonomy, see Prichard, *Researches*, pp. 80–5.
96. Lawrence, *Natural History of Man*, p. 269.
97. Ibid., p. 269.
98. Ibid.
99. Ibid.

100. Ibid., p. 282.
101. Ibid., p. 283.
102. Ibid. Lawrence's opinion of beauty is representative of early nineteenth-century Eurocentric conceptions.
103. Ibid.
104. Ibid.
105. Ibid.
106. J. R. Forster, *The Resolution Journal of Johann Reinhold Forster*, ed. M. Hoare, 4 vols (London: Hakluyt Society, 1982), vol. 2, pp. 298–9.
107. There is substantial secondary literature on the link between ethnography and voyages of exploration. For examinations on the impact of the Cook voyages on the study of human diversity, see: G. Obeyesekere, *The Apotheosis of Captain Cook: European Mythmaking in the Pacific* (Princeton, NJ: Princeton University Press, 1992); J. Stagl, *A History of Curiosity: The Theory of Travel 1550–1800* (Chur: Harwood Academic Publishers, 1995); A. Calder, J. Lamb and B. Orr (eds), *Voyages and Beaches: Pacific Encounters, 1769–1840* (Honolulu, HI: University of Hawaii Press, 1999); R. Love, *Maritime Exploration in the Age of Discovery, 1415–1800* (London: Greenwood Publishing Group, 2006); and G. Russell, 'An "Entertainment of Oddities": Fashionable Sociability and the Pacific in the 1770s', in K. Wilson (ed.), *A New Imperial History: Culture, Identity and Modernity in Britain and the Empire 1660–1840* (Cambridge: Cambridge University Press, 2008), pp. 48–70.
108. For a complete list of the sources and figures Prichard used in the first edition of the *Researches*, see Stocking, 'Guide to Prichard's References'.
109. Prichard, *Researches*, pp. 243–4.
110. Ibid., p. 244.
111. Ibid., p. 245.
112. Ibid.
113. M. T. Bravo, 'Ethnological Encounters', in N. Jardine, J. A. Secord and E. C. Spary (eds), *Cultures of Natural History* (Cambridge: Cambridge University Press, 1996), pp. 338–57, on p. 339.
114. Prichard, *Researches*, p. 279.
115. Ibid., p. 247.
116. Ibid., p. 248.
117. Ibid., p. 249.
118. Ibid.
119. Augstein, *James Cowles Prichard's Anthropology*, p. 10; and S. Qureshi, *Peoples on Parade: Exhibitions, Empire and Anthropology in Nineteenth-Century Britain* (Chicago, IL: University of Chicago Press, 2011), pp. 269–70.
120. Prichard, *Researches*, p. 282.
121. Ibid., p. 251.
122. Prichard, *Researches*, p. 251. Prichard was also referring to Jean-Marie de Gérando's description of a 'philosophical traveller', which was essentially a trained observer of human diversities. See J. M. de Gérando, *Considérations sur les Diverses Méthodes a Suivre dans l'Observation des Peuples Sauvages* (Paris: Societé des Observateurs de l'Homme, 1800).
123. Prichard, *Researches*, p. 251.
124. Ibid.
125. Ibid., pp. 251–2.
126. Ibid., p. 253.
127. Ibid.

128. Ibid., p. 255.
129. Prichard, *Researches*, pp. 256–7. Prichard relied heavily on Dampier's four-volume travelogue *Voyages Around the World* from 1729. For more, see W. Dampier, *A Collection of Voyages*, 4 vols (London: James and John Knapton, 1729).
130. Prichard, *Researches*, p. 257. Prichard used Bougainville's *Voyage Autour du Monde* from 1771. For more, see L. A. Bougainville, *Voyage Autour du Monde: Par le Frégate du Roi la Boudeuse, et la Flûte l'*Étoile (Paris: Saillant and Nyon, 1771).
131. Prichard, *Researches*, pp. 264–5; Dampier, *A Collection of Voyages*, vol. 1, p. 464.
132. Prichard, *Researches*, p. 267.
133. Lawrence, *Natural History of Man*, pp. 107–8.
134. Ibid., p. 108.
135. Ibid.
136. Ibid., pp. 108–9.
137. Ibid., p. 109.
138. Ibid., p. 109–10.
139. Ibid., p. 215.
140. For more on botanical exchange networks in Britain and the colonies, see: A. Secord, 'Corresponding Interests: Artisans and Gentlemen in Nineteenth-Century Natural History', *British Journal for the History of Science*, 27 (1994), pp. 383–408; D. P. Miller, *Visions of Empire: Voyages, Botany, and Representations of Nature* (Cambridge: Cambridge University Press, 1996); and J. Endersby, *Imperial Nature: Joseph Hooker and the Practice of Victorian Science* (Chicago, IL: University of Chicago Press, 2008).
141. Hodgkin, 'Obituary of Dr. Prichard', p. 182.
142. Ibid., p. 192.
143. Unlike Lawrence's *Natural History of Man*, which was reprinted eight more times as new editions, each of Prichard's three editions of the *Researches* were markedly different. For more, see J. C. Prichard, *Researches into the Physical History of Mankind*, 2nd edn, 2 vols (London: John and Arthur Arch, 1826); and J. C. Prichard, *Researches into the Physical History of Mankind*, 3rd edn, 5 vols (London: Sherwood, Gilbert and Piper, 1836–47). In 1851 Houlston and Stoneman published a fourth edition of Prichard's *Researches*, which was a reprint of the third edition; see J. C. Prichard, *Researches into the Physical History of Mankind*, 4th edn, 5 vols (London: Houlston, and Stoneman, 1851–1858). Prichard's work was translated into other languages: for a German translation, see J. C. Prichard, *Naturgeschichte des Menschengeschletchs*, trans. R. Wagner and J. G. F. Will, 5 vols (Leipzig: Verlag von Leopold Vos., 1840–48); for a French translation, see J. C. Prichard, *Histoire Naturelle de l'Homme*, trans. F. Roulin (Paris: J. Baillière, 1843).
144. Carpenter, 'Art V.-1 Researches into the Physical History of Mankind', p. 429.

## 2 Ethnology in Transit: Informants, Questionnaires and the Formation of the Ethnological Society of London

1. R. King, 'Address to the Ethnological Society of London Delivered at the Anniversary, 25th May 1844', *Journal of the Ethnological Society of London*, 2 (1850), pp. 9–42, on p. 19.
2. For background information on the history of anthropology prior to the nineteenth century, see H. Liebersohn, 'Anthropology before Anthropology', in H. Kuklick (ed.), *A New History of Anthropology* (Oxford: Blackwell Publishing, 2008), pp. 17–31.
3. Lawrence, *Natural History of Man*, p. 242.

4. See: Bravo, 'Ethnological Encounters'; Kenny, 'From the Curse of Ham to the Curse of Nature'; Kuklick, 'The British Tradition'; Stocking, 'What's in a Name?'; and Urry, 'Notes and Queries on Anthropology', pp. 45–57. For more information on colonial networks of exchange, see Z. Laidlaw, *Colonial Connections 1815–1845: Patronage, the Information Revolution and Colonial Government* (Manchester: Manchester University Press, 2005), pp. 13–35.
5. Bravo, 'Ethnological Encounters', p. 342.
6. Secord, 'Corresponding Interests'; Miller, *Visions of Empire*; and Endersby, *Imperial Nature*.
7. J. M. de Gérando, *Considerations on the Diverse Methods to Follow in the Observations of Savage Peoples*, trans. F. C. T. Moore (London: Routledge & Kegan Paul, 1969), p. 70.
8. G. Stocking, 'French Anthropology in 1800', *Isis*, 55 (1964), pp. 134–50, on p. 134.
9. R. Parkin, 'The French-Speaking Countries', in F. Barth, A. Gingrich, R. Parkin and S. Silverman, *One Discipline Four Ways: British, German, French and American Anthropology* (Chicago, IL: University of Chicago Press, 2005), pp. 157–253, on p. 157.
10. For examples, see: Stocking, 'French Anthropology in 1800'; Parkin, 'The French-Speaking Countries'; and E. Sibeud, 'The Metamorphosis of Ethnology in France, 1839–1930', in H. Kuklick (ed.), *A New History of Anthropology* (Oxford: Blackwell Publishing, 2008), pp. 96–110.
11. Sibeud, 'The Metamorphosis of Ethnology in France', pp. 97–100. For a history of the Société des Observateurs de l'Homme, see J. L. Chappey, *La Société des Observateurs de l'Homme (1799–1804)* (Paris: Société des Études Robespierristes, 2002); and Stocking, 'French Anthropology in 1800'.
12. Bravo, 'Ethnological Encounters', pp. 342–7.
13. P. Rabinow, *French Modern: Norms and Forms of the Social Environment* (Cambridge, MA: MIT Press, 1989), p. 19.
14. Ibid., pp. 20–1.
15. Ibid., p. 21.
16. A reproduction of Cuvier's *Note Instructive* is located in G. Cuvier, 'Note Instructive sur les Recherches à Faire Relativement aux Différences Anatomiques des Diverses Races d'Hommes', in J. Copans and J. Jamin (eds), *Aux Origines de l'Anthropologie Française: Les Mémoires de la Société des Observateurs de l'Homme en l'An VIII* (Paris: Sycomore, 1978), pp. 171–6.
17. For the original French edition, see de Gérando, *Considérations sur les Diverses Méthodes a Suivre*. An English translation of the guide was produced in 1969; see de Gérando, *Considerations on the Diverse Methods to Follow*. See also Rabinow, *French Modern*, pp. 21–4.
18. Rabinow, *French Modern*, pp. 20–1. For more information on Baudin's expedition, see F. Péron, *Voyage de Découverte aux Terres Australes* (Paris: Arthus Bertrand Libraire Éditeur, 1824).
19. De Gérando, *Considerations on the Diverse Methods to Follow*, p. 20.
20. For more information on the history of the Royal Asiatic Society, see S. Digby and S. Simmonds (eds), *The Royal Asiatic Society: Its History and Treasures* (Leiden: Brill, 1979).
21. For more information on the Aborigines Protection Society, see Stocking, 'What's in a Name?', pp. 369–71; Cantor, *Quakers, Jews and Science*, pp. 133–8; Desmond and Moore, *Darwin's Sacred Cause*, pp. 143–4; and J. Heartfield, *The Aborigines' Protection Society: Humanitarian Imperialism in Australia, New Zealand, Fiji, Canada, South Africa, and the Congo, 1836–1909* (New York: Columbia University Press, 2011). Richard King also discussed the history of the science of Man in Britain in his 1844

presidential address at the Ethnological Society of London; see King, 'Address to the Ethnological Society of London'.

22. Kenny, 'From the Curse of Ham to the Curse of Nature', p. 369.
23. Initially, the ESL held weekly meetings at Hodgkin's home; however, as the society continued to grow, it relocated to larger locations. For more on Hodgkin's early abolitionist activities, see Z. Laidlaw, 'Heathens, Slaves and Aborigines: Thomas Hodgkin's Critique of Missions and Anti-slavery', *History Workshop Journal*, 64 (2007), pp. 134–61.
24. For more information on the Ethnological Society of London, see: Stocking, 'What's in a Name?'; Stocking, *Victorian Anthropology*, pp. 238–45; Kenny, 'From the Curse of Ham to the Curse of Nature'; Kuklick, 'The British Tradition'.
25. Anon., 'Regulations', *Journal of the Ethnological Society of London*, 1 (1848), pp. 3–14, on p. 3.
26. Prichard, *Researches*; Lawrence, *Natural History of Man.*
27. Stocking, *Victorian Anthropology*, p. 243.
28. For a complete list of Prichard's published works, see Stocking, 'Guide to Prichard's References'.
29. For information on the circulation of knowledge through print, see: J. Topham, 'Scientific Publishing and the Reading of Science in Nineteenth-Century Britain: A Historiographical Survey and Guide to Sources', *Studies in History and Philosophy of Science*, 31 (2000), pp. 559–612; G. Cantor and S. Shuttlesworth, 'Introduction', in G. Cantor and S. Shuttleworth (eds), *Science Serialized: Representation of the Sciences in Nineteenth-Century Periodicals* (Cambridge, MA: MIT Press, 2004), pp.1–15; G. Dawson, R. Noakes and J. Topham, 'Introduction', in G. Cantor et al., *Science in the Nineteenth-Century Periodical: Reading the Magazine of Nature* (Cambridge: Cambridge University Press, 2004), pp. 1–34; and J. Secord, 'Knowledge in Transit', *Isis*, 95 (2004), pp. 654–72.
30. King, 'Address to the Ethnological Society of London', p. 9.
31. Ibid., p. 15.
32. Carpenter, 'Art V.-1 Researches into the Physical History of Mankind', pp. 429–30. For example, see the five-volume third edition of Prichard's book, *Researches*.
33. For more on botanical exchange networks in Britain and the colonies, see Secord, 'Corresponding Interests'; Miller, *Visions of Empire*; and Endersby, *Imperial Nature*.
34. For more on the relationship between Darwin's research and his correspondences, see J. Browne, *Charles Darwin: The Power of Place* (New York: Alfred A. Knopf, 2002).
35. King, 'Address to the Ethnological Society of London', p. 15.
36. For more information on George Back and Richard King's Arctic Expedition, see J. Cavell, *Tracing the Connected Narrative: Artic Exploration in British Print Culture, 1818–1860* (Toronto: University of Toronto Press, 2008).
37. For biographical information on King, see E. Baigent, 'King, Richard (1810/11–1876), Artic Traveller and Ethnologist', *ODNB*; and Stocking, 'What's in a Name?', p. 372.
38. For biographical information on Hodgkin, see: A. M. Kass, 'Hodgkin, Thomas (1798–1866), Physician and Social Reformer', *ODNB*; A. M. Kass and E. H. Kass, *Perfecting the World: The Life and Times of Dr. Thomas Hodgkin 1798–1866* (Boston, MA: Harcourt Brace Jovanovich, 1988); and Laidlaw, 'Heathens, Slaves and Aborigines'.
39. For information on Crawfurd, see C. M. Turnbull, 'Crawfurd, John (1783–1868), Orientalist and Colonial Administrator', *ODNB*; and Stocking, 'What's in a Name?', p. 375.
40. See T. Hodgkin, 'The Progress of Ethnology', *Journal of the Ethnological Society of London*, 1 (1848), pp. 27–45; and E. Dieffenbach, 'The Study of Ethnology', *Journal of the Ethnological Society of London*, 1 (1848), pp. 15–26.

41. See R. G. Latham, 'On the Languages of the Oregon Territory', *Journal of the Ethnological Society of London*, 1 (1848), pp. 154–66; and R. G. Latham, 'On the Ethnography of Russian America', *Journal of the Ethnological Society of London*, 1 (1848), pp. 182–91.
42. For biographical information on Latham, see A. M. Ridler, 'Latham, Robert Gordon (1812–1888), Ethnologist and Philologist', *ODNB*. See also Kenny, 'From the Curse of Ham to the Curse of Nature'; and Stocking, 'What's in a Name?'
43. R. King, 'On the Physical Characters of the Esquimaux', *Journal of the Ethnological Society of London*, 1 (1848), pp. 45–59; R. King, 'On the Intellectual Character of the Esquimaux', *Journal of the Ethnological Society of London*, 1 (1848), pp. 127–53; R. King, 'On the Industrial Arts of the Esquimaux', *Journal of the Ethnological Society of London*, 1 (1848), pp. 277–300; and J. Crawfurd, 'On the Malayan and Polynesian Languages and Races', *Journal of the Ethnological Society of London*, 1 (1848), pp. 330–74.
44. For examples of Prichard's use of Forster's observations, see Prichard, *Researches*, p. 282.
45. Lawrence, *Natural History of Man*, p. 108.
46. King, 'Address to the Ethnological Society of London', p. 15.
47. For more on colonial rule and ethnology, see: H. Kuklick, *Imperial Bureaucrat: Colonial Administrative Service in the Gold Coast 1920–1939* (Stanford, CA: Hoover Institution Press, 1979); T. Miller, *Colonizing Egypt* (Cambridge: Cambridge University Press, 1988); and Laidlaw, *Colonial Connections 1815–1845*.
48. King, 'Address to the Ethnological Society of London', p. 16.
49. For more, see the first edition of Prichard's work: Prichard, *Researches*, p. 244.
50. Lawrence, *Natural History of Man*, p. 242.
51. King, 'Address to the Ethnological Society of London', p. 16.
52. J. C. Prichard, 'Anniversary Address for 1848, to the Ethnological Society of London on the Recent Progress of Ethnology', *Journal of the Ethnological Society of London*, 2 (1850), pp. 119–49, on p. 120.
53. King, 'Address to the Ethnological Society of London', p. 9.
54. Ibid., p. 16.
55. Ibid., p. 20.
56. Minutes from the Proceedings of the Ethnological Society of London Meeting, 22 June 1844, Royal Anthropological Institute, ESL Minutes, London (hereafter ESL Minutes), A1, p. 10. In 1845 the minutes discuss that the ESL was to send a copy of its journal to the Geographical Statistical Society of Darmstadt and the Philosophical Society of London. See ESL Minutes, 22 November 1845, p. 22.
57. Dieffenbach, 'The Study of Ethnology', p. 16.
58. Ibid.
59. Ibid.
60. Ibid., pp. 16–21.
61. Ibid., p. 18.
62. Ibid., p. 19.
63. Martineau, *How to Observe*, p. 1.
64. See J. C. Prichard et al., 'Queries Respecting the Human Race to be Addressed to Travellers and Others', in Anon. (ed.), *Report of the Eleventh Meeting of the British Association for the Advancement of Science* (London: John Murray, 1842), pp. 332–9; and Prichard, 'Ethnology'. Prichard wrote the questionnaire in 1848 prior to his death, and it was released the following year. See also Bravo, 'Ethnological Encounters', pp. 342–7.
65. Parkin, 'The French-Speaking Countries', p. 160. Parkin was referring to Montesquieu as being the first French writer to consider ethnological and anthropological questions.

However, an examination of Montesquieu's contributions to the study of human diversity is far too encompassing for a discussion here.

66. Hodgkin and Cull, 'A Manual of Ethnological Inquiry', p. 194.
67. Ibid., p. 193. Unfortunately, Hodgkin and Cull did not provide their readers with a reference or publication date for the original guidebook designed by the Paris Ethnological Society. However, Hodgkin likely read this questionnaire when he visited the Paris Society in 1839. See Stocking, 'What's in a Name?', p. 371; and Stocking, 'French Anthropology in 1800'.
68. Desmond and Moore, *Darwin's Sacred Cause*, p. 151.
69. Hodgkin and Cull, 'A Manual of Ethnological Inquiry', p. 193.
70. For de Gérando, the *voyageur philosophe* was a special informant trained to observe and document extra-European people *in situ*, utilizing the ethnographic framework outlined in his *Considérations*. See de Gérando, *Considerations on the Diverse Methods to Follow*, pp. 20, 70.
71. Cuvier, 'Note Instructive', pp. 171–6.
72. See Rabinow, *French Modern*, p. 22; and de Gérando, *Considerations on the Diverse Methods to Follow*, pp. 63–70.
73. De Gérando, *Considerations on the Diverse Methods to Follow*, p. 65.
74. Ibid., p. 66.
75. Ibid.
76. Ibid., p. 67.
77. Ibid.
78. Ibid., p. 70.
79. See the first edition of Prichard's book, *Researches*.
80. De Gérando, *Considerations on the Diverse Methods to Follow*, p. 68.
81. Ibid., p. 75.
82. Ibid., p. 78.
83. Ibid.
84. Lawrence, *Natural History of Man*, p. 217.
85. De Gérando, *Considerations on the Diverse Methods to Follow*, p. 78.
86. Ibid.
87. Ibid., pp. 79–80.
88. Ibid., p. 88.
89. Ibid., p. 89.
90. Obeyesekere, *The Apotheosis of Captain Cook*, pp. 3–7.
91. De Gérando, *Considerations on the Diverse Methods to Follow*, p. 94.
92. Martineau, *How to Observe*, p. 2.
93. Ibid., p. 4.
94. Ibid.
95. Prichard et al., 'Queries Respecting the Human Race', p. 332. For more information on the BAAS committee and the handbook, see Desmond and Moore, *Darwin's Sacred Cause*, pp. 151–2.
96. Prichard et al., 'Queries Respecting the Human Race'.
97. Prichard, 'Ethnology'.
98. For more, see Secord, 'Corresponding Interests'; and Endersby, *Imperial Nature*, pp. 31–4, 54–9.

99. T. Wright and J. C. Prichard, 'Ethnology', in J. F. W. Herschel (ed.), *A Manual of Scientific Enquiry; Prepared for the Use of Officers in Her Majesty's Navy and Travellers in General* (London: John Murray, 1859), pp. 253–67, on p. 253.
100. Ibid., p. 253.
101. For the table of contents of the ethnology chapter, see J. F. W. Herschel (ed.), *A Manual of Scientific Enquiry; Prepared for the Use of Officers in Her Majesty's Navy and Travellers in General* (London: John Murray, 1859), p. xv.
102. Herschel, *A Manual of Scientific Enquiry*, p. xv; and Wright and Prichard, 'Ethnology', pp. 253–67.
103. Hodgkin and Cull, 'A Manual of Ethnological Inquiry', p. 193.
104. Ibid., p. 194.
105. Ibid., p. 193. For more information on John Arthur Phillips, see J. Morrell, *John Phillips and the Business of Victorian Science* (Aldershot: Ashgate, 2005).
106. Hodgkin and Cull, 'A Manual of Ethnological Inquiry', p. 194.
107. Ibid.
108. See the first edition of Prichard's book, *Researches*, pp. 55–8; and Lawrence, *Natural History of Man*, pp. 282–4.
109. R. Cull, 'Remarks on the Nature, Objects, and Evidences of Ethnological Science, an Address Read at a Conversazione on June 4, 1851', *Journal of the Ethnological Society of London*, 2 (1854), pp. 103–11, on p. 103.
110. Hodgkin and Cull, 'A Manual of Ethnological Inquiry', p. 197.
111. J. C. Prichard, *The Eastern Origin of Celtic Nations Proved by a Comparison to Sanskrit, Greek, Latin and Teutonic Languages* (London: John and Arthur Arch, 1831). Hodgkin also discusses 'language' as a valuable source of ethnological evidence in Hodgkin, 'The Progress of Ethnology', pp. 27–8.
112. Hodgkin and Cull, 'A Manual of Ethnological Inquiry', p. 200.
113. Ibid., pp. 200–3.
114. Ibid., p. 201.
115. For more on colonial administration and control, see Kuklick, *Imperial Bureaucrat*; Miller, *Colonizing Egypt*; and Laidlaw, *Colonial Connections 1815–1845*.
116. Hodgkin and Cull, 'A Manual of Ethnological Inquiry', pp. 203–4.
117. Ibid., p. 203.
118. Ibid., pp. 204–8.
119. Kuklick, 'The British Tradition', p. 54; Urry, 'Notes and Queries on Anthropology', pp. 45–57.
120. Stocking, *Victorian Anthropology*, p. 259; Kuklick, 'The British Tradition, p. 56. See also: E. H. Man, 'On the Aboriginal Inhabitants of the Andaman Islands (Part I)', *Journal of the Anthropological Institute of Great Britain and Ireland*, 12 (1882), pp. 69–116; E. H. Man, 'On the Aboriginal Inhabitants of the Andaman Islands (Part II)', *Journal of the Anthropological Institute of Great Britain and Ireland*, 12 (1882), pp. 117–75; and E. H. Man, 'On the Aboriginal Inhabitants of the Andaman Islands (Part III)', *Journal of the Anthropological Institute of Great Britain and Ireland*, 12 (1882), pp. 327–434.
121. Cull, 'Remarks on the Nature', p. 113.
122. C. Malcolm, 'Address to the Ethnological Society of London, Delivered at the Anniversary, 14th May, 1851', *Journal of the Ethnological Society of London*, 3 (1854), pp. 86–102, on p. 86.
123. Ibid., p. 87.

124. For a comprehensive bibliography of Brodie, see W. H. Brock, 'Brodie, Sir. Benjamin Collins, First Baronet (1783–1862), Physiologist and Surgeon', *ODNB*.
125. R. Knox, *The Races of Man: A Fragment* (London: Henry Renshaw, 1850).

## 3 Ethnology at Home: Robert Gordon Latham, Robert Knox and Competing Observational Practices

1. Cull, 'Remarks on the Nature', p. 103.
2. There were variations of both monogenism and polygenism. However, for my purposes, I use these terms strictly to demarcate those who argued in favour of the single origin of humans in contrast to those that argued in favour of the multiple origins of humans.
3. As discussed in Chapter 2, there were other ethnologists producing monogenetic models, and figures such as Hodgkin and King were key developers of this theoretical paradigm.
4. For more on Étienne Geoffroy Saint-Hilaire and his morphology, see T. Appel, *The Cuvier-Geoffroy Debate: French Biology in the Decades before Darwin*, (Oxford: Oxford University Press, 1987), pp. 69–104.
5. For more on the British ethnological community during the 1850s, see: Stocking, 'What's in a Name?'; Stocking, *Victorian Anthropology*, pp. 238–45; Kenny, 'From the Curse of Ham to the Curse of Nature'; and Kuklick, 'The British Tradition', pp. 52–6.
6. For the most comprehensive discussion on Knox's extreme views, consult: Stocking, *Victorian Anthropology*, pp. 65–9, 245–7; E. Richards, 'The "Moral Anatomy" of Robert Knox: The Interplay between Biological and Social Thought in Victorian Scientific Naturalism', *Journal of the History of Biology*, 22 (1989), pp. 373–436; Efron, *Defenders of Race*, pp. 45–57; Kenny, 'From the Curse of Ham to the Curse of Nature'; Desmond and Moore, *Darwin's Sacred Cause*, pp. 38–45, 103–4, 188–91, 304–6, 335–41, 381–3; and A. W. Bates, *The Anatomy of Robert Knox: Murder, Mad Science and Medical Regulation in Nineteenth-Century Edinburgh* (Eastbourne: Sussex Academic Press, 2010).
7. I recognize that terms such as Gypsy/Gypsies and Bushman are contentious; however, I use them here because these terms were part of Latham's and Knox's taxonomical vocabulary.
8. Anon., 'The Varieties of Man', *Bentley's Miscellany*, 28 (1850), pp. 393–5, on p. 393.
9. For more bibliographic information on Latham, see: Anon., 'Robert Gordon Latham, M.D., F.R.S.', in L. Reeve (ed.), *Portraits of Men of Eminence in Literature, Science and Art, with Biographical Memoirs*, 6 vols (London: Lovell Reeve & Co., 1863), vol. 1, pp. 93–7; Stocking, *Victorian Anthropology*, pp. 100–9; Ridler, 'Latham, Robert Gordon', *ODNB*; and E. Sera-Shriar, 'Ethnology in the Metropole: Robert Knox, Robert Gordon Latham and Local Sites of Observational Training', *Studies in History and Philosophy of Biological and Biomedical Sciences*, 42 (2011), pp. 486–96, on pp. 491–5. Latham was also a key reformer of the English language, and he produced several grammar books and a dictionary. See: R. G. Latham, *An Elementary English Grammar, for the Use of Schools* (London: Taylor, Walton and Maberly, 1843); R. G. Latham, *History and Etymology of the English Language* (London: Taylor, Walton and Maberly, 1849); R. G. Latham, *A Hand-Book of the English Language: For the Use of Students of the Universities and Higher Classes of School* (London: Taylor, Walton and Maberly, 1851); R. G. Latham, *Elements of Comparative Philology* (London: Taylor, Walton and Maberly, 1862); R. G. Latham, *A Dictionary of the English Language*, 2 vols (London: Longmans and Green, 1864–70); and R. G. Latham, *Essential Rules and Principles for the Study of English Grammar* (London: Longmans and Green, 1876).

10. Metagraphy is the process of transliterating the letters from one language's alphabet into its equivalent in another.
11. Anon., 'Robert Gordon Latham, M.D., F.R.S.', p. 93.
12. For example, see: R. G. Latham, *The Ethnology of Europe* (London: John Van Voorst, 1852), pp. 191–220; R. G. Latham, *Descriptive Ethnology*, 2 vols (London: John Van Voorst, 1859), vol. 1, pp. 421–5; and Latham, *Descriptive Ethnology*, vol. 2, pp. 36–40.
13. Ridler, 'Latham, Robert Gordon'; and Anon., 'Robert Gordon Latham, M.D., F.R.S.', p. 93.
14. For Latham's main ethnological works, see: Latham, 'On the Ethnography of Russian America'; R. G. Latham, *The Natural History of the Varieties of Man* (London: John Van Voorst, 1850); R. G. Latham, 'General and Special Apophthegms', *Edinburgh New Philosophical Journal*, 51 (1851), pp. 39–44; Latham, *The Ethnology of Europe*; R. G. Latham, 'Ethnology', in E. Forbes and R. G. Latham, *The Natural History Department of the Crystal Palace Described* (London: Crystal Palace Library, 1854), pp. 5–79; vols 1–2 of Latham, *Descriptive Ethnology*; and R. G. Latham, 'The Antiquity of the Human Race', *Westminster Review*, 74 (1860), pp. 411–36.
15. Latham, *Varieties of Man*, p. vii. Stocking also argued in his book *Victorian Anthropology* that Latham was a key figure to forward Prichardian ethnological principles after Prichard's death in 1848; see p. 53.
16. Latham, *Varieties of Man*, p. vii.
17. Ibid., p. 1.
18. Ibid., pp. 2–3. Latham also refers to the 'Mongolian' variety as the 'Mongolidae', the 'Negro' variety as both the 'African' and the 'Atlantidae', and the 'European' variety as the 'Iapetidae'.
19. Ibid., p. 2.
20. Blumenbach, *Anthropological Treatises*, pp. 237–9; Prichard, *Researches*, pp. 46, 55–8; Lawrence, *Natural History of Man*, pp. 236–52.
21. Latham, *Varieties of Man*, pp. 3–4.
22. Ibid., p. 4.
23. Ibid.
24. Ibid., p. 7.
25. Ibid.
26. Ibid., p. 8.
27. Ibid.
28. Ibid.
29. Ibid.
30. Ibid., pp. 8–9.
31. Ibid., p. 9.
32. Ibid.
33. Ibid., p. 10.
34. Van Diemen's Land was renamed Tasmania in 1856 to honour the explorer Abel Tasman, who was the first European to land on its shores in 1643.
35. Latham, *Varieties of Man*, p. 244.
36. Ibid., p. 245–46.
37. G. Blainey, *A Land Half Won* (London: Macmillan Press, 1980), p. 75.
38. We know less about William Hare; he was born between 1792 and 1804, and after the execution of William Burke, Hare left Edinburgh and disappeared from historical record, so it is uncertain when he died.

39. The William Burke and William Hare murders (1827–8) occurred in a neighbourhood in Edinburgh known as West Port. For details of the Burke and Hare murders, see: Anon., *An Account of the Life of William Burke, who was Executed on Wednesday the 28th Day of January, 1829, for Murder: With the Confession He Made while Under Sentence of Death* (Dunfermline: John Miller, 1829); Anon., 'A Sketch of the Life and Writings of Robert Knox, the Anatomist', *Examiner*, 3279 (1870), pp. 773–4; I. Rae, *Knox the Anatomist* (Edinburgh: Oliver and Boyd, 1964), pp. 122–54; H. Lonsdale, *A Sketch of the Life and Writings of Robert Knox, the Anatomist* (London: Macmillan, 1870), pp. 73–89; and Bates, *The Anatomy of Robert Knox.*
40. Knox's followers regularly promoted his importance for science and medicine. For examples, see C. C. Blake, 'The Life of Robert Knox, the Anatomist. A Sketch of his Life and Writings by Henry Lonsdale', *Journal of Anthropology*, 1 (1871), pp. 332–8; and J. Hunt, 'On the Origin of the Anthropological Review and its Connection with the Anthropological Society', *Anthropological Review*, 6 (1868), pp. 431–42.
41. Richards, 'The "Moral Anatomy" of Robert Knox', p. 377.
42. For bibliographic information on Robert Knox, consult: Lonsdale, *A Sketch of the Life and Writings of Robert Knox*; Rae, *Knox the Anatomist*; C. L. Taylor, 'Knox, Robert (1791–1862), Anatomist and Ethnologist', *ODNB*; and H. J. Blackman, 'Knox, Robert, (1793–1862)', in B. Lightman (ed.), *Dictionary of Nineteenth-Century British Scientists*, 4 vols (London: Thoemmes Continuum, 2004), vol. 3, pp. 1160–2. For larger examinations on Knox's career as a physician and anatomist, see Richards, 'The "Moral Anatomy" of Robert Knox'; and R. Richardson, *Death Dissection and the Destitute* (London: Routledge & Kegan Paul, 1987), pp. 132–46.
43. Taylor, 'Robert Knox', *ODNB*.
44. For more information on extramural medical education in Edinburgh, see Jacyna, *Philosophical Whigs*, pp. 78–92.
45. Rae, *Knox the Anatomist*, pp. 14–15; and Lonsdale, *A Sketch of the Life and Writings of Robert Knox*, p. 13.
46. Lonsdale, *A Sketch of the Life and Writings of Robert Knox*, pp. 11–13.
47. Ibid., p. 11.
48. Ibid., p. 12.
49. Ibid., p. 13.
50. For example, see Knox, *Races of Man*, pp. 156–7.
51. Ibid., p. 7.
52. Ibid.
53. Ibid., p. 8.
54. Ibid., p. 9.
55. Ibid.
56. R. Knox, *Great Artists and Great Anatomists* (London: John Van Voorst, 1852), pp. 1–5.
57. Ibid., p. 2.
58. R. Knox, *A Manual of Human Anatomy* (London: Henry Renshaw, 1853), p. vii.
59. See Knox, *Races of Man*, p. 20. For more on Saint-Hilaire and transcendentalism's influence on Knox's ethnology, see Knox, *Great Artists and Great Anatomists*, pp. 2–4.
60. Knox, *Races of Man*, p. 9.
61. Knox, *Great Artists and Great Anatomists*, pp. 6–7.
62. Knox, *Races of Man*, pp. 9–10.
63. Ibid., p. 11.
64. Ibid., p. 19.

65. Ibid.
66. Ibid., p. 12.
67. Ibid., p. 13.
68. Latham, 'Ethnology', p. 6.
69. Latham, 'Ethnology', pp. 5–79. For more information on the Great Exhibition, see G. Cantor, *Religion and the Great Exhibition of 1851* (Oxford: Oxford University Press, 2012). For more on the Sydenham site, see J. Schwartz, 'Un Beau Souvenir du Canada', in E. Edwards and J. Hart (eds), *Photographs Objects Histories: On the Materiality of Images* (New York: Routledge, 2004), pp. 16–32, on pp. 26–7.
70. E. Forbes and R. G. Latham, *The Natural History Department of the Crystal Palace Described* (London: Crystal Palace Library, 1854).
71. Latham, 'Ethnology', p. 6. For more on the ethnological exhibit see ch. 6 of Qureshi, *Peoples on Parade*, pp. 185–221. See also S. Qureshi, 'Robert Gordon Latham, Displayed Peoples and the Natural History of Race, 1854–1866', *Historical Journal*, 54 (2011), pp. 143–66.
72. J. Conolly, *The Ethnological Exhibitions of London* (London: John Churchill, 1855), pp. 41–2.
73. J. Hunt, 'Anthropological News', *Anthropological Review*, 5 (1867), pp. 240–56, on pp. 241–2. There was a second fire at the Crystal Palace in 1936. However, by this time most of the original ethnological exhibit had long been destroyed.
74. Latham, 'Ethnology', p. 6.
75. Originally Latham intended to have nude mannequins, but after hearing objections from church officials, he opted to clothe adult figures in customary dress. For more, see Anon., 'The Crystal Palace at Sydenham', *Daily News*, Friday, 9 June 1854.
76. Latham, 'Ethnology', p. 75. Sir Robert Hermann Schomburgk (1804–65) was an explorer, botanist and surveyor for the British military and lived for several years in British Guiana. Thus he had a strong familiarity of the people and culture of that region.
77. Ibid., p. 57.
78. Ibid.
79. Ibid., p. 58.
80. Ibid.
81. The missionary museum in London also had displays that were designed to showcase extra-European cultural practices. For more, see S. Sivasundaram, *Nature and the Godly Empire: Science and Evangelical Mission in the Pacific 1795–1850* (Cambridge: Cambridge University Press, 2005), pp. 185–9.
82. Latham, 'Ethnology', p. 54.
83. For more information about these images at the Pitt Rivers Museum, see Qureshi, *Peoples on Parade*, pp. 194–8; and Qureshi, 'Robert Gordon Latham'. See also: 'Bushman Display from the Crystal Palace Exhibition', Pitt Rivers Museum, Photographic Collections Catalogue, Oxford (hereafter Pitt Rivers Coll.), PRM 1998.211.9; and 'Zulu Display from the Crystal Palace Exhibition', Pitt Rivers Coll., PRM 1998.210.3.
84. Lonsdale, *A Sketch of the Life and Writings of Robert Knox*, p. 331.
85. Knox, *Races of Man*, p. 10.
86. The most notable examples were Prichard, *Researches*, vol. 3; and Latham, *The Ethnology of Europe*.
87. Lonsdale, *A Sketch of the Life and Writings of Robert Knox*, p. 292.

88. Knox, *Races of Man*, p. 103. For more on the historical representation of Gypsy communities, see: B. Vesey-FitzGerald, *Gypsies of Britain: An Introduction to their History* (Newton Abbot: David and Charles, 1973); J. Okely, *The Traveller-Gypsies* (Cambridge: Cambridge University Press, 1983), pp. 1–25; and D. E. Nord, *Gypsies and the British Imagination, 1807–1930* (New York: Columbia University Press, 2008).
89. Knox, *Races of Man*, p. 108.
90. Ibid., p. 109.
91. Ibid.
92. Ibid., p. 109.
93. Ibid., p. 131.
94. Ibid., p. 133.
95. Ibid., pp. 135–7.
96. Ibid., p. 137.
97. Ibid.
98. Blake, 'The Life of Robert Knox, the Anatomist', p. 336.
99. Anon., 'The Natural History of the Varieties of Man', *Examiner*, 2230 (1850), pp. 686–7, on p. 686.
100. Ibid., p. 687.
101. Anon., 'Art II. – The Natural History of the Varieties of Man', *Prospective Review*, 24 (1850), pp. 449–58, on p. 449.
102. Ibid., p. 449.
103. Ibid., p. 457.
104. Prichard, *Natural History of Man*.
105. Anon., 'The Natural History of the Varieties of Man', *Athenaeum*, 1226 (1851), pp. 447–8, on p. 447.
106. Ibid.
107. Ibid.
108. Malcolm, 'Address to the Ethnological Society of London', p. 91.
109. Ibid.
110. Ibid.
111. For examples see: Anon., 'The Varieties of Man'; J. Mitford, 'The Races of Men: A Fragment. By Robert Knox M.D.', *Gentleman's Magazine* (November 1850), pp. 528–9; Anon., 'The Races of Mankind', *Fraser's Magazine for Town and Country*, 44 (1851), pp. 654–65; and Anon., 'Great Artists and Great Anatomists', *Athenaeum*, 1297 (1852), pp. 935–7.
112. Mitford, 'The Races of Men', p. 528.
113. Ibid., p. 529.
114. Anon., 'Ethnology', *Literary Gazette*, 1765 (1850), pp. 855–7, on p. 857.
115. Ibid., p. 857.
116. Ibid.
117. Anon., 'The Varieties of Man', pp. 393–4.
118. Ibid., p. 394.
119. Ibid.
120. Anon., 'The Races of Mankind', p. 655.
121. Ibid., p. 656.
122. Ibid.
123. Ibid.
124. Anon., 'Great Artists and Great Anatomists', p. 935.
125. Ibid.

126. Ibid.
127. Ibid., p. 937.
128. J. Hunt, 'Introductory Address on the Study of Anthropology', *Anthropological Review*, 1 (1863), pp. 1–20, on p. 8.
129. Ibid.

## 4 The Battle for Mankind: James Hunt, Thomas Huxley and the Emergence of British Anthropology

1. Hunt, 'Introductory Address on the Study of Anthropology', p. 4.
2. For more on the British ethnological community during the 1850s, see: Stocking 'What's in a Name?'; Stocking, *Victorian Anthropology*, pp. 238–45; Kenny, 'From the Curse of Ham to the Curse of Nature'; Kuklick 'The British Tradition', pp. 52–6.
3. I recognized that the term 'racially sensitive' is anachronistic. However, it is instructive, and I want to emphasize how the historiography has positioned members of the ESL in opposition to members at the ASL. The terms 'racist' and 'sexist' are also contentious because nineteenth-century European ideas about extra-Europeans and gender roles vary substantially from twentieth- and twenty-first-century understandings. For an account of Hunt and his views on slavery and race, consult D. Lorimer, *Colour, Class and the Victorians: English Attitudes to the Negro in the Mid-Nineteenth-Century* (Leicester: Leicester University Press, 1978). For an account of Hunt's and Huxley's views on women and societies, consult E. Richards, 'Huxley and Woman's Place in Science: The Woman Question and the Control of Victorian Anthropology', in J. Moore (ed.), *History, Humanity and Evolution* (Cambridge: Cambridge University Press, 1989), pp. 253–84.
4. Stocking, *Victorian Anthropology*, p. 249.
5. For more on the relationship between nineteenth-century gender politics and anthropology, see E. Fee, 'The Sexual Politics of Victorian Social Anthropology', *Feminist Studies*, 1 (1973), pp. 23–39.
6. Stocking, 'What's in a Name?', p. 376.
7. For historiography on Hunt, see: J. W. Burrow, 'Evolution and Anthropology in the 1860's: The Anthropological Society of London, 1863–71', *Victorian Studies*, 7 (1963), pp. 137–49; Stocking, 'What's in a Name?'; Kuklick, 'The British Tradition'; Kenny, 'From the Curse of Ham to the Curse of Nature'; F. Spencer, 'Hunt, James (1833–1869)', in F. Spencer (ed.), *History of Physical Anthropology*, 2 vols (London: Garland Publishing, Inc., 1997), vol. 1, pp. 506–8.
8. In the first half of the nineteenth century, an early treatment for 'stammering' and 'stuttering' involved a series of painful injections into the patient's throat. See J. Hunt, *Stammering and Stuttering, their Nature and Treatment* (London: Longman, Green Longman and Roberts, 1861), pp. 157–60.
9. D. Lorimer, 'Hunt, James 1833–1869', in B. Lightman (ed.), *Dictionary of Nineteenth-Century British Scientists*, 4 vols (London: Thoemmes Continuum, 2004), vol. 2, pp. 1033–5, on p. 1034.
10. The degree to which Hunt dominated the ASL was not entirely realized until 1868, after an exchange of letters was published in the *Athenaeum*. Hunt further declared his control of the anthropological community in an 1868 article from his self-financed journal the *Anthropological Review*. For more, see: H. Clarke, 'Anthropological Society of London', *Athenaeum*, 2129 (1868), p. 210; H. Clarke, 'Anthropological Society of London',

*Athenaeum*, 2131 (1868), pp. 271–2; H. Clarke, 'Anthropological Society of London', *Athenaeum*, 2143 (1868), pp. 681–2; H. Clarke and H. Brookes, 'Anthropological Society of London', *Athenaeum*, 2134 (1868), pp. 368–9; H. Brookes and E. W. Brabrook, 'Anthropological Society of London', *Athenaeum*, 2132 (1868), p. 302; C. Harding and D. I. Heath, 'Anthropological Society of London', *Athenaeum*, 2133 (1868), p. 334; J. Hunt, 'Anthropological Society of London', *Athenaeum*, 2135 (1868), p. 402; J. Hunt and J. F. Collingwood, 'Anthropological Society of London', *Athenaeum*, 2130 (1868), pp. 239–40; and Hunt, 'On the Origin of the Anthropological Review', pp. 431–2.

11. For examples, see: Hunt, 'Introductory Address on the Study of Anthropology'; J. Hunt, 'On Physio-Anthropology, its Aim and Method', *Journal of the Anthropological Society of London*, 5 (1867), pp. ccix–cclxxi; and J. Hunt, *On the Negro's Place in Nature* (London: Trübner and Co., 1863).
12. Hunt, 'On the Origin of the Anthropological Review', pp. 432–4.
13. The best examples of Hunt's manipulation of his periodicals are from his reports of the proceedings at the BAAS. These reports were anonymous and were highly critical of the leading members of the ethnological community at the BAAS. See: Anon., 'Anthropology at the British Association', *Anthropological Review*, 1 (1863), pp. 379–464; Anon., 'Anthropology at the British Association, A.D. 1864', *Anthropological Review*, 2 (1864), pp. 294–335; Anon., 'On the Prospects of Anthropological Science at the British Association of 1865', *Anthropological Review*, 3 (1865), 224–9.
14. J. Hunt, 'Abstract of a Report of the Ethnological Papers Read at the Manchester Meeting of the British Association in September 1861', *Transactions of the Ethnological Society of London*, 2 (1863), pp. 1–3, on p. 2.
15. King, 'Address to the Ethnological Society of London'; Hodgkin and Cull, 'A Manual of Ethnological Inquiry'.
16. Hunt, 'Abstract of a Report of the Ethnological Papers', p. 2.
17. Lorimer, 'Hunt, James 1833–1869', p. 1034. For several examples of Hunt's criticisms of ethnology, see Hunt, 'Introductory Address on the Study of Anthropology'; and Hunt, 'On the Origin of the Anthropological Review'.
18. Wallace wrote a short article in 1864 where he outlined some of the differences between anthropological and ethnological theories and methods. For more, see A. R. Wallace, 'On the Origin of Human Races and the Antiquity of Man Deduced from the Theory of Natural Selection', *Journal of the Anthropological Society of London*, 2 (1864), pp. clviii–clxxxvii.
19. Stocking, *Victorian Anthropology*, p. 252. For more on the X-Club, see: R. Macleod, 'The X Club a Social Network of Science in Late-Victorian England', *Notes and Records of the Royal Society of London*, 24 (1970), pp. 305–22; F. Turner, 'The Victorian Conflict between Science and Religion: A Professional Dimension', *Isis*, 69 (1978), pp. 356–76; R. Barton, 'Huxley, Lubbock, and Half a Dozen Others', *Isis*, 89 (1998), pp. 410–44; and A. Desmond, 'Redefining the X Axis: "Professionals", "Amateurs" and the Making of Mid-Victorian Biology – A Progress Report', *Journal of the History of Biology*, 34 (2001), pp. 3–50.
20. T. H. Huxley to L. Cooke, 4 May 1864, located in L. Huxley (ed.), *Life and Letters of Thomas Henry Huxley*, 2 vols (London: Macmillan and Co., 1900), vol. 1, p. 251.
21. S. Lyons, *Thomas Henry Huxley: The Evolution of a Scientist* (Amherst, NY: Prometheus Books, 1999), p. 25. For more information on Thomas Henry Huxley, see the following biographies: W. Irvine, *Thomas Henry Huxley* (London: Longmans, 1960); C. Bibby, *Scientist Extraordinary: The Life and Scientific Work of Thomas Henry Huxley 1825–1895* (Oxford: Pergamon, 1972); M. DiGregorio, *T. H. Huxley's Place in Natu-*

*ral Science* (New Haven, CT: Yale University Press, 1984); J. V. Jensen, *Thomas Henry Huxley: Communicating for Science* (Newark, DE: University of Delaware Press, 1991); A. Desmond, *Huxley: The Devil's Disciple* (London: Michael Joseph, 1994); and A. Desmond, *Huxley: Evolution's High Priest* (London: Michael Joseph, 1997).

22. T. H. Huxley, *Evidence as to Man's Place in Nature* (London: Williams and Norgate, 1863); and Huxley, 'On the Methods and Results of Ethnology'.
23. Lyons, *Thomas Henry Huxley*, pp. 27–8.
24. Kuklick, *The Savage Within*, pp. 43–4; and J. Huxley (ed.), *T. H. Huxley's Diary of the Voyage of H.M.S. Rattlesnake* (London: Chatto and Windus, 1935), pp. 154–71.
25. For more on Huxley's early career, see B. Lightman, *Victorian Popularizers of Science: Designing Nature for New Audiences* (Chicago, IL: University of Chicago Press, 2007), pp. 357–8.
26. Lyons, *Thomas Henry Huxley*, pp. 29–34; Kuklick, *The Savage Within*, pp. 43–4; and Stocking, *Victorian Anthropology*, pp. 249–57.
27. Huxley, *Life and Letters of Thomas Henry Huxley*, vol. 1, pp. 249, 262.
28. Huxley, *Evidence as to Man's Place in Nature*; and Huxley, 'On the Methods and Results of Ethnology'.
29. Huxley's drive to place all the scientific disciplines onto an evolutionary framework is far too big a subject for a discussion here. For more on Huxley and evolutionary theory, see DiGregorio, *T. H. Huxley's Place in Natural Science*, pp. 53–67.
30. For the relationship between ethnology and Darwinism and the debates between monogenists and polygenists, see ch. 8 of Desmond and Moore, *Darwin's Sacred Cause*, pp. 199–227.
31. For more on the debates relating to politics, religion and human origin in the nineteenth century, see D. Livingstone, *Adam's Ancestors: Race, Religion and the Politics of Human Origins* (Baltimore, MD: Johns Hopkins University Press, 2008).
32. M. Fichman, 'Biology and Politics: Defining the Boundaries', in B. Lightman (ed.), *Victorian Science in Context* (Chicago, IL: University of Chicago Press, 1997), pp. 94–118, on pp. 94–101.
33. For more on Huxley's vocational strategy, see: E. Caudill, 'The Bishop-Eaters: The Publicity Campaign for Darwin and on the Origin of Species', *Journal of the History of Ideas*, 55 (1994), pp. 441–60; P. White, *Thomas Huxley: Making the 'Man of Science'* (Cambridge: Cambridge University Press, 2003), pp. 51–8; and J. Elwick, *Styles of Reasoning in the British Life Sciences: Shared Assumptions, 1820–1858* (London: Pickering & Chatto, 2007), pp. 131–59. Lightman has also discussed Huxley's vocational strategy in *Victorian Popularizers of Science*, p. 359–61.
34. Richards, 'Huxley and Woman's Place in Science', p. 267.
35. Ibid., pp. 262–7.
36. Stocking, *Victorian Anthropology*, pp. 252–4. For one of the best examples of Hunt attacking Huxley and the other scientific naturalists, see J. Hunt, 'On the Application of the Principle of Natural Selection to Anthropology, in Reply to the Views Advocated by Some of Mr Darwin's Disciples', *Anthropological Review*, 4 (1866), pp. 320–40.
37. T. H. Huxley, 'Emancipation – Black and White', *Reader*, 5 (1865), pp. 561–2, on p. 561.
38. Hunt, 'On the Origin of the Anthropological Review', p. 433.
39. J. Hunt, 'Dedication to Dr. Paul Broca', in C. Vogt, *Lectures on Man: His Place in Creation, and in the History of the Earth*, ed. J. Hunt (London: Longman, Green, Longman and Roberts, 1864), pp. v–x, on p. viii.

40. Hunt, 'Dedication to Dr. Paul Broca', p. viii; and Hunt, 'On the Origin of the Anthropological Review', p. 434. For more on Hunt's views on women at ESL meetings, see Richards, 'Huxley and Woman's Place in Science', pp. 254–7.
41. Hunt responded to the critiques of his opponents in his 1868 article; see Hunt, 'On the Origin of the Anthropological Review'.
42. See Minutes from the Proceedings of the Anthropological Society of London Meetings 1863–71, 5 August 1863, Royal Anthropological Institute, ASL Minute Book, London (hereafter ASL Minutes), A3; and Richards, 'Huxley and Woman's Place in Science', p. 263.
43. Huxley, 'Emancipation – Black and White', p. 561.
44. See A. R. Wallace to T. H. Huxley, 26 February 1864, Thomas Henry Huxley Papers, College Archives, Imperial College London, London (hereafter Huxley Papers), ICL. HP 28.91.
45. Anon., 'Front Matters', *Journal of the Ethnological Society of London*, 1 (1869), vi–xxviii, on p. xiv.
46. Ibid.
47. For more information on Victorians and race, consult C. Hall, K. McClelland and J. Rendall, *Defining the Victorian Nation: Class, Race, Gender and the Reform Act of 1867* (Cambridge: Cambridge University Press, 2000); and J. Haller, 'Concepts of Race Inferiority in Nineteenth-Century Anthropology', *Journal of the History of Medicine*, 25 (1970), pp. 40–51.
48. See 13 March 1861, ESL Minutes. The newly revamped periodical for the Ethnological Society of London was renamed the *Transactions of the Ethnological Society of London*.
49. Stocking, *Victorian Anthropology*, p. 247; 4 March 1862, ESL Minutes; and 5 May 1863, ESL Minutes.
50. The best example of Hunt's views on race was published in his first major contribution to anthropology, entitled *On the Negro's Place in Nature*.
51. Hunt, 'On the Application of the Principle of Natural Selection to Anthropology', p. 336. For more on Knox's racial determinism, see Knox, *Races of Man*, pp. 1–7.
52. For Hunt's critiques of Darwinian evolutionary theory, see J. Hunt, 'On the Doctrine of Continuity Applied to Anthropology', *Anthropological Review*, 5 (1867), pp. 110–20; and Hunt, 'On the Application of the Principle of Natural Selection to Anthropology'.
53. Hunt, *On the Negro's Place in Nature*, p. 4.
54. Ibid., p. 27.
55. Stocking, *Victorian Anthropology*, pp. 247–8.
56. Knox argued in his *Races of Man* that Africans were unable to progress to the same level as Europeans. See Knox, *Races of Man*.
57. R. Clarke, 'Sketches of the Colony of Sierra Leone and its Inhabitants', *Transactions of the Ethnological Society of London*, 2 (1863), pp. 320–63, on p. 329.
58. Ibid., p. 329.
59. Hunt, *On the Negro's Place in Nature*, p. 27.
60. Unfortunately Robert Clarke does not give his wife Mrs Clarke's given name.
61. For more information on nineteenth-century visual representations of Africans, consult: P. Brantlinger, 'Victorians and Africans: The Genealogy of the Myth of the Dark Continent', *Critical Inquiry*, 12 (1985), pp. 166-203; P. D. Curtin, *The Image of Africa: British Ideas and Action, 1780–1850* (Madison, WI: University of Wisconsin Press, 1964); and D. Bindman, *Ape to Apollo: Aesthetics and the Idea of Race in the 18th Century* (Ithaca, NY: Cornell University Press, 2002), pp. 23–78.
62. Clarke, 'Sketches of the Colony at Sierra Leone', p. 348.

63. 5 May 1863, ESL Minutes; and Stocking, 'What's in a Name?', p. 376.
64. T. Wright, 'Preface', *Transactions of the Ethnological Society*, 2 (1863), p. iii–viii, on p. v.
65. Ibid., p. v.
66. See T. H. Huxley to C. Darwin, Darwin Correspondence, Letter number 4228, 2 July 1863, at http://www.darwinproject.ac.uk/entry-4228 [accessed 8 August 2012].
67. Anon., 'Back Matters', *Transactions of the Ethnological Society*, 2 (1863), pp. 1–10 (back pages), on p. 8. The ESL lists Hunt as an honorary member right up until 1867; see Anon., 'Back Matters', *Transactions of the Ethnological Society*, 5 (1867), pp. 1–12 (back pages), on p. 10.
68. Kenny, 'From the Curse of Ham to the Curse of Nature', p. 337; Stocking, 'What's in a Name?', pp. 375–81; and Burrow, 'Evolution and Anthropology in the 1860's', pp. 137–49.
69. See T. H. Huxley's letter to L. Cooke, 4 May 1864, located in Huxley, *Life and Letters of Thomas Henry Huxley*, vol. 1, pp. 251–2.
70. Huxley, 'Emancipation – Black and White', p. 561.
71. Huxley, 'On the Methods and Results of Ethnology', p. 257.
72. Ibid., pp. 257–9.
73. Hunt, *On the Negro's Place in Nature*, p. 3.
74. Hunt, 'Introductory Address on the Study of Anthropology', p. 2.
75. Ibid.
76. Knox, *Races of Man*, p. 9; and Knox, *A Manual of Human Anatomy*, p. vii.
77. Hunt, 'On Physio-Anthropology, its Aim and Method', p. ccx.
78. Ibid., p. ccxi.
79. Ibid., p. ccxii.
80. Ibid.
81. M. Rudwick, *The Great Devonian Controversy: The Shaping of Scientific Knowledge among Gentlemanly Specialists* (Chicago, IL: University of Chicago Press, 1985), pp. 24–5. For more on Baconianism and disciplinary formation, see: R. Porter, *The Making of Geology: Earth Science in Britain 1660–1815* (Cambridge: Cambridge University Press, 1977), pp. 66–70; J. Morrell, 'The Legacy of William Smith: The Case of John Phillips in the 1820s', *Archives of Natural History*, 16 (1989), pp. 319–35; A. B. Van Riper, *Men among the Mammoths* (Chicago, IL: University of Chicago, 1993), pp. 15–44; P. Findlen, *Possessing Nature: Museums, Collecting and Scientific Culture in Early Modern Italy*, (Berkeley, CA: University of California Press, 1994), pp. 400–1; and P. J. A. Levine, *The Amateur and the Professional: Antiquarians, Historians and Archaeologists in Victorian England 1838–1886* (Cambridge: Cambridge University Press, 2003).
82. Hunt, 'On Physio-Anthropology, its Aim and Method', p. ccxii.
83. Ibid., pp. ccxii–ccxiii. For more on rhetorical strategies in science between 1830 and 1917, see R. Yeo, 'Scientific Method and the Rhetoric of Science in Britain 1830–1917', in J. Schuster and R. Yeo (eds), *The Politics and Rhetoric of Scientific Method* (Dordrecht: D. Reidel Publishing Company, 1986), pp. 259–97.
84. Hunt, *On the Negro's Place in Nature*, p. 4.
85. Ibid., p. 6.
86. Desmond and Moore, *Darwin's Sacred Cause*, pp. 199–227.
87. J. Hunt, 'Address Delivered at the Third Anniversary Meeting of the Anthropological Society of London', *Journal of the Anthropological Society of London*, 4 (1866), pp. lix–lxxxi, on pp. lxv–lxvi.

88. For more on the North American school of anthropology see S. Silverman, 'The United States', in F. Barth, A. Gingrich, R. Parkin and S. Silverman, *One Discipline Four Ways: British, German, French and American Anthropology* (Chicago, IL: University of Chicago Press, 2005), pp. 257–347; and R. Darnell, 'North American Traditions in Anthropology: The Historiographic Baseline', in H. Kuklick (ed.), *A New History of Anthropology* (Oxford: Blackwell Publishing, 2008), pp. 35–51.
89. Hunt, 'On the Doctrine of Continuity Applied to Anthropology', pp. 111–12.
90. Hunt, 'On the Doctrine of Continuity Applied to Anthropology', p. 113. Darwin does not discuss human evolution until the publication of the *Descent of Man* in 1871. In it, he responds to some of the criticisms levelled at him by the anthropologicals. See C. Darwin, *The Descent of Man, and Selection in Relation to Sex*, 2 vols (London: John Murray, 1871) vol. 1, p. 228.
91. Hunt, 'On the Doctrine of Continuity Applied to Anthropology', p. 113.
92. Ibid., p. 114.
93. Hunt attacked the theory of 'natural selection' in his response to Huxley's article on 'On the Methods and Results of Ethnology'. See Hunt, 'On the Application of the Principle of Natural Selection to Anthropology'; and Huxley, 'On the Methods and Results of Ethnology'.
94. Hunt, 'On the Doctrine of Continuity Applied to Anthropology', pp. 117–18.
95. White, *Thomas Huxley*, p. 54; and M. Bartholomew, 'Huxley's Defence of Darwin', *Annals of Science*, 32 (1975) 525–35.
96. White, *Thomas Huxley*, p. 56. For more on Huxley's institutional reforms, see also: S. Forgan, 'The Architecture of Display: Museums, Universities and Objects in Nineteenth-Century Britain', *History of Science*, 32 (1994), pp. 139–62; G. Gooday, '"A Fungoid Assemblage of Buildings": Diversity and Adversity in the Development of College Architecture and Scientific Education in Nineteenth-Century South Kensington', *History of Universities*, 13 (1994), pp. 153–92; and G. Gooday, 'Constructing South Kensington: The Buildings and Politics of T. H. Huxley's Working Environments', *British Journal for the History of Science*, 29 (1996), pp. 435–68.
97. Huxley, 'On the Methods and Results of Ethnology', p. 257.
98. Ibid., pp. 257–8.
99. Ibid., p. 257.
100. Hunt, 'On the Application of the Principle of Natural Selection to Anthropology', pp. 322–3.
101. Huxley, 'On the Methods and Results of Ethnology', p. 260.
102. Ibid., p. 263.
103. Ibid., pp. 263–4.
104. Ibid., p. 263.
105. Ibid., p. 263–4.
106. Ibid., p. 269.
107. Ibid.
108. For instance, see Prichard, *Researches*, pp. 17–24.
109. Huxley, 'On the Methods and Results of Ethnology', pp. 269–70.
110. M. Dixon and G. Radick, *Darwin in Ilkley* (Stroud: History Press, 2009), p. 101. For further readings on Huxley and natural selection, see: T. H. Huxley to C. Darwin, 6 May 1862, located in Huxley, *Life and Letters of Thomas Henry Huxley*, vol. 1, pp. 205–6; Bartholomew, 'Huxley's Defence of Darwin', pp. 525–35; G. Radick, *The Simian Tongue: The Long Debate about Animal Language* (Chicago, IL: University of Chicago Press, 2007), pp. 32–3; and Appendix VI in F. Burkhardt, D. Porter, J. Harvey and J. Topham,

*The Correspondence of Charles Darwin*, 18 vols (Cambridge: Cambridge University Press, 1985– ), vol. 10, pp. 700–11.

111. T. H. Huxley, 'Time and Life: Mr. Darwin's "Origin of Species"', *Macmillan's Magazine*, 2 (1859), pp. 142–8, on p. 148. See also B. Lightman, *The Origins of Agnosticism: Victorian Unbelief and the Limits of Knowledge* (Baltimore, MD: Johns Hopkins University Press, 1987), pp. 156–7.
112. Hunt, 'Introductory Address on the Study of Anthropology', p. 4.
113. Ibid., p. 11.
114. Ibid.
115. Hodgkin and Cull, 'A Manual of Ethnological Inquiry'; and King, 'Address to the Ethnological Society of London'.
116. Hunt, 'Introductory Address on the Study of Anthropology', p. 12.
117. Ibid.
118. Ibid., p. 13.
119. J. Hunt, 'The President's Address', *Journal of the Ethnological Society of London*, 2 (1864), pp. lxxx–xcv, on p. lxxx.
120. Ibid., p. lxxx.
121. J. Hunt, 'The President's Address', *Journal of the Anthropological Society of London*, 5 (1867), pp. xliv–lxxi, on p. li.
122. Ibid., p. lvi.
123. Hunt, 'Introductory Address on the Study of Anthropology', p. 15.
124. P. Broca, *On the Phenomena of Hybridity in the Genus Homo*, trans. C. C. Blake (London: Longman, Green Longman and Roberts, 1864); G. Pouchet, *On the Plurality of the Human Race*, trans. H. Beaven (London: Longman, Green, Longman and Roberts, 1864); C. Vogt, *Lectures on Man: His Place in Creation, and in the History of the Earth*, ed. J. Hunt (London: Longman, Green, Longman and Roberts, 1864).
125. T. H. Huxley to J. Fayrer, 14 June 1866, located in Huxley, *Life and Letters of Thomas Henry Huxley*, vol. 1, p. 274. See ch. 6 of E. Edwards, *Raw Histories: Photographs, Anthropology and Museums* (Oxford: Berg, 2001), pp. 131–55.
126. T. H. Huxley to J. Fayrer, 14 June 1866, located in Huxley, *Life and Letters of Thomas Henry Huxley*, vol. 1, p. 274.
127. Ibid.
128. See Huxley Papers, ICL.HP.30.f.75.
129. Edwards, *Raw Histories*, pp. 131–2.
130. The original notice can be found in the Public Record Office in London: T. H. Huxley to the Colonial Office of London, 30 November 1869, National Archives, London, Public Record Office (hereafter PRO London), PRO.CO.854/10.
131. This image was photographed by two colonial informants named William Lawrence (1835–1905) and David McKenzie Selkirk (1829–1904) during the early 1870s. It depicts a thirty-two-year-old convict stationed at the Breakwater Prison in Cape Town, South Africa. It is from the Pitt-Rivers photographic collection in Oxford; see 'Full-length and full face anthropometric photograph of Abraham Lucas', Pitt Rivers Coll., PRM 1998.211.13.2; and 'Full-length profile anthropometric photograph of Abraham Lucas', Pitt Rivers Coll., PRM 1998.211.13.3. For more on Lawrence and Selkirk and mid-nineteenth-century anthropometric photography in South Africa, see C. Webster, 'The Portrait Cabinet of Dr. Bleek: Anthropometric Photographs by Early Cape Photographers', *Critical Arts*, 14 (2000), pp. 1–15. For more on anthropology and photography generally, see: E. Edwards, 'Photographic Types: The Pursuit of Method', *Visual*

*Anthropology*, 3 (1990), pp. 235–58; T. Wright, 'Photography: Theories of Realism and Convention', in E. Edwards (ed.), *Anthropology and Photography 1860–1920* (New Haven, CT: Yale University Press, 1992), pp. 18–31; J. Scherer, 'The Photographic Document: Photographs as Primary Data in Anthropological Enquiry', in E. Edwards (ed.), *Anthropology and Photography 1860–1920* (New Haven, CT: Yale University Press, 1992), pp. 32–41; and E. Edwards and C. Morton, 'Introduction', in E. Edwards and C. Morton (eds), *Photography, Anthropology and History: Expanding the Frame* (Farnham: Ashgate, 2009), pp. 1–24.

132. Edwards, *Raw Histories*, pp. 137–8.
133. Ibid., p. 138.
134. The original letter can be found in the Public Record Office in London: A. Musgrave to T. H. Huxley, PRO London, PRO.CO.60/38/332.
135. For more, see Webster, 'The Portrait Cabinet of Dr. Bleek'.
136. J. H. Lamprey and H. Clark, 'Ethnological Notes and Queries', *Journal of the Ethnological Society of London*, 1 (1869), pp. 84–8, on pp. 84–5.
137. T. H. Huxley, 'On the Ethnology and Archaeology of India: Opening Address of the President', *Journal of Ethnological Society of London*, 1 (1869), pp. 89–93, on pp. 89–90.
138. Anon., 'Death of the Best Man in England', *Anthropological Review*, 8 (1870), p. 97.
139. E. S. Collingwood, 'Death of Dr. James Hunt: Science or Sun-Stroke', *British Medical Journal*, 2 (1869), p. 355.
140. Ibid.
141. For more on Huxley's efforts to amalgamate ethnology and anthropology at the end of the 1860s, see: Stocking, 'What's in a Name?'; Kenny, 'From the Curse of Ham to the Curse of Nature'; and E. Sera-Shriar, 'Observing Human Difference: James Hunt, Thomas Huxley and Competing Disciplinary Strategies in the 1860s', *Annals of Science*, forthcoming.

## 5 Synthesizing the Discipline: Charles Darwin, Edward Burnett Tylor and Developmental Anthropology in the Early 1870s

1. Darwin, *Descent of Man*, vol. 1, p. 1.
2. For more on the anthropological schism, see: Burrow, 'Evolution and Anthropology in the 1860's'; Stocking, 'What's in a Name?'; Stocking, *Victorian Anthropology*, pp. 238–45; Kenny, 'From the Curse of Ham to the Curse of Nature'; and Kuklick, 'The British Tradition', pp. 52–6.
3. For more on Darwin, race questions and human origins, see D. Paul, 'Darwin, Social Darwinism and Eugenics', in J. Hodge and G. Radick (eds), *The Cambridge Companion to Darwin* (Cambridge: Cambridge University Press, 2003), pp. 214–40, on pp. 223–4; and Desmond and Moore, *Darwin's Sacred Cause*. For an example of Hunt's criticisms of Darwinian theory, see Hunt, 'On the Application of the Principle of Natural Selection to Anthropology'; and Hunt, 'On the Doctrine of Continuity Applied to Anthropology'.
4. G. Stocking, *Delimiting Anthropology: Occasional Essays and Reflections* (Madison, WI: University of Wisconsin Press, 2001), p. 105.
5. Stocking, *Victorian Anthropology*, pp. 238–73; G. Stocking, *After Tylor: British Social Anthropology 1888-1951* (Madison, WI: University of Wisconsin Press, 1995); Stocking, *Delimiting Anthropology*, pp. 103–12; Barth, 'Britain and the Commonwealth', pp. 3–10; and Kuklick, 'The British Tradition', pp. 52–70.

6. G. Stocking, 'The Ethnographer's Magic: Fieldwork in British Anthropology from Tylor to Malinowski', in G. Stocking, *The Ethnographer's Magic and Other Essays in the History of Anthropology* (Madison, WI: University of Wisconsin Press, 1992), pp. 12–59, on p. 19.
7. E. B. Tylor, *Anahuac: or Mexico and the Mexicans, Ancient and Modern* (London: Longman, Green, Longman and Roberts, 1861), p. 1.
8. Stocking, *Victorian Anthropology*, pp. 144–5. For a more thorough discussion on the impact of Darwin's theories on anthropology, see ch. 12 of Desmond and Moore, *Darwin's Sacred Cause*, pp. 317–47.
9. Desmond and Moore, *Darwin*, p. 22.
10. For more on medicine at Edinburgh, see: Augstein, *James Cowles Prichard's Anthropology*, pp. 3–5; Jacyna, *Philosophical Whigs*, pp. 85–92; Rosner, *Medical Education in the Age of Improvement*; Desmond and Moore, *Darwin*, pp. 21–31; and Desmond and Moore, *Darwin's Sacred Cause*, pp. 27–48.
11. Desmond and Moore, *Darwin's Sacred Cause*, pp. 32–7.
12. D. Dean, 'Jameson, Robert (1774–1854), Geologist and Natural Historian', *ODNB*.
13. Desmond and Moore, *Darwin's Sacred Cause*, p. 28.
14. Ibid., p. 29.
15. Ibid., p. 28.
16. Ibid., pp. 49–67.
17. See R. Fitzroy and P. P. King, *Narrative of the Surveying Voyages of His Majesty's Ships Adventure and Beagle, between the Years 1826–1836, Describing their Examination of the southern Shores of South America, and the Beagle's Circumnavigation of the Globe*, 3 vols (London: Henry Colburn, 1839) vol. 2, p. 18; and R. D. Keynes, 'Introduction', in C. Darwin, *Beagle Diary*, ed. R. D. Keynes (Cambridge: Cambridge University Press, 2001), pp. xi–xxiv, on p. xii.
18. Paul, 'Darwin, Social Darwinism and Eugenics', p. 218.
19. Desmond and Moore, *Darwin's Sacred Cause*, p. 12.
20. From Darwin's diary, 20 January 1831. See C. Darwin, *Beagle Diary*, ed. R. D. Keynes (Cambridge: Cambridge University Press, 2001), p. 26; and H. Gruber, *Darwin on Man: A Psychological Study of Scientific Creativity* (London: Wildwood House London, 1974), p. 181.
21. From Darwin's diary, 26 November 1834; see Darwin, *Beagle Diary*, p. 266.
22. From Darwin's diary, 22 December 1835; see ibid., p. 384.
23. From Darwin's diary, 22 December 1835; see ibid.
24. Lawrence, *Natural History of Man*, p. 106.
25. From Darwin's diary, 22 December 1835; see Darwin, *Beagle Diary*, p. 385.
26. From Darwin's diary, 18 December 1832; see ibid., p. 122; Gruber, *Darwin on Man*, p. 182; Desmond and Moore, *Darwin*, p. 132–3; and Desmond and Moore, *Darwin's Sacred Cause*, pp. 68–110, G. Radick, 'Did Darwin Change his Mind about the Fuegians?', *Endeavour*, 34 (2010), pp. 50–4.
27. Paul, 'Darwin, Social Darwinism and Eugenics', p. 218.
28. Darwin, 'Transmution of Species', Notebook C, p. 204; and Desmond and Moore, *Darwin's Sacred Cause*, pp. 157–8.
29. Prichard et al., 'Queries Respecting the Human Race'.
30. For more on Darwin's health problems, see Dixon and Radick, *Darwin in Ilkley*.
31. For example, see Stocking, *After Tylor*, pp. 3–4.

32. A. Lang, 'Edward Burnett Tylor', in H. Balfour (ed.), *Anthropological Essays Presented to Edward Burnett Tylor in Honour of his 75th Birthday* (Oxford: Clarendon Press, 1907), pp. 1–17, on p. 1.
33. E. Isichei, *Victorian Quakers* (Oxford: Oxford University Press, 1970), p. 7; and Cantor, *Quakers, Jews, and Science*, pp. 234–5.
34. For more, see the first edition of Prichard, *Researches*, pp. 3–4.
35. W. H. George, 'Alfred Tylor (1824–1884), Geologist and Brass-Founder', *ODNB*.
36. Stocking, *Victorian Anthropology*, p. 157.
37. Tylor, *Anahuac*, p. 1.
38. Stocking, *Victorian Anthropology*, p. 157.
39. Tylor, *Anahuac*, p. iii.
40. Ibid., p. 147.
41. Ibid.
42. Ibid., p. 148.
43. Ibid., p. 23.
44. Stocking, 'What's in Name?'; and Kenny, 'From the Curse of Ham to the Curse of Nature'.
45. Stocking, *Victorian Anthropology*, p. 159.
46. E. B. Tylor, *Researches into the Early History of Mankind and the Development of Civilisation* (London: John Murray, 1865). For more, see Stocking, *Delimiting Anthropology*, p. 108; and J. Leopold, *Culture in Comparative and Evolutionary Perspective: E. B. Tylor and the Making of Primitive Culture* (Berlin: Dietrich Reimer Verlag, 1980).
47. Darwin, *Descent of Man*, vol. 1, p. 1.
48. Huxley, 'Time and Life: Mr. Darwin's "Origin of Species"', p. 148.
49. Darwin, *Descent of Man*, vol. 1, pp. 1–2.
50. C. Darwin, *The Origin of Species by Means of Natural Selection or the Preservation of Favoured Races in the Struggle for Life* (London: John Murray, 1859), p. 6.
51. Darwin, *Descent of Man*, vol. 1, pp. 2–3.
52. Huxley, 'On the Methods and Results of Ethnology', p. 257.
53. For examples of ethnological classification systems, see: Prichard, *Researches*; Prichard, *Researches*, 2nd edn, vols 1–2; Lawrence, *Natural History of Man*; Prichard, *Researches*, 3rd edn, vols 1–5; Latham, *Varieties of Man*; and Huxley, 'On the Methods and Results of Ethnology'.
54. Darwin, *Descent of Man*, vol. 1, p. 3. See also Desmond and Moore, *Darwin's Sacred Cause*, pp. 283–4.
55. Desmond and Moore, *Darwin's Sacred Cause*, p. 158.
56. Hunt was one of Darwin's main opponents within British anthropological circles during the 1860s. For more, see Hunt, 'Introductory Address on the Study of Anthropology', pp. 2–3 and 8–9. See also Hunt, 'On the Application of the Principle of Natural Selection to Anthropology'; and Hunt, 'On the Doctrine of Continuity Applied to Anthropology'.
57. Darwin, *Descent of Man*, vol. 1, p. 3.
58. Ibid., pp. 9–157.
59. Ibid., p. 214.
60. Ibid.
61. Ibid.
62. See for example Hunt, 'On the Application of the Principle of Natural Selection to Anthropology', pp. 329–32.

63. For instance, Prichard discusses sterility in relation to species and varieties in the introduction of the first edition to his *Researches*; see pp. 9–10.
64. Darwin, *Descent of Man*, vol. 1, pp. 214–15.
65. Ibid.
66. Ibid., p. 215.
67. Ibid., p. 217.
68. Ibid., p. 215.
69. Ibid., pp. 215–16. See also Lawrence, *Natural History of Man*, p. 106; and Darwin, 'Transmutation of Species', Notebook C, p. 204.
70. Darwin, *Descent of Man*, vol. 1, p. 216.
71. Ibid., pp. 225–6.
72. Ibid., p. 226.
73. Hunt, 'Introductory Address on the Study of Anthropology', pp. 4–7.
74. Darwin, *Descent of Man*, vol. 1, p. 225.
75. Ibid., p. 226.
76. Ibid., pp. 227–8.
77. Prichard, *Researches*, pp. 25–32; and Prichard, *Researches*, 3rd edn, vol. 4, pp. 454, 519, 534, 537. See also Desmond and Moore, *Darwin's Sacred Cause*, p. 284.
78. Latham, *Varieties of Man*, p. 8.
79. Darwin discusses at length his 'principle of sexual selection' in chapter VIII *Descent of Man*, vol. 1, pp. 253–320.
80. Darwin, *Descent of Man*, vol. 2, p. 341.
81. Ibid., p. 355.
82. E. B. Tylor, *Primitive Culture: Researches into the Development of Mythology, Philosophy, Religion, Language, Art and Custom*, 2 vols (London: John Murray, 1871), vol. 1, p. v. Unless otherwise noted, all subsequent citations are to this edition.
83. Tylor called this his theory of 'animism', but a detailed discussion of this aspect of his work is far too comprehensive for an analysis here. His 'animism' theory formed the basis of the second volume of *Primitive Culture*.
84. Tylor, *Primitive Culture*, vol. 1, p. vi.
85. Ibid., pp. v–vi.
86. For example, see Hunt, 'Introductory Address on the Study of Anthropology', p. 4; Hunt, 'On Physio-Anthropology, its Aim and Method', pp. ccx–ccxi; and Huxley Papers, ICL.HP.30.f.75.
87. Tylor, *Primitive Culture*, vol. 1, p. vi.
88. Ibid.
89. J. Crawfurd, *A Grammar and Dictionary of the Malay Language, with a Preliminary Dissertation*, 2 vols (London: Smith, Elder and Co., 1852).
90. Tylor's notebooks are undated and poorly catalogued, with no official Pitt-Rivers Museum number. The above material is located in E. B. Tylor, Notebooks, 55 vols, MS Oxford, Pitt Rivers Museum (hereafter Tylor Notebooks), vol. 1, p. 51.
91. E. B. Tylor, *Primitive Culture: Researches into the Development of Mythology, Philosophy, Religion, Language, Art and Custom*, 2nd edn, 2 vols (London Henry Holt and Company, 1889), vol. 1, pp. 1, 26–30.
92. Stocking, *Delimiting Anthropology*, pp. 109–10.
93. Tylor, *Primitive Culture*, vol. 1, p. 1.
94. Ibid., p. 2.
95. Ibid.

96. Ibid.
97. For more, see Darwin's discussion on the 'causes of variability' in *Origin of Species*, pp. 7–43.
98. See the second edition of Tylor, *Primitive Culture*, vol. 1, p. 3.
99. Ibid.
100. Prichard, *Researches*, p. 7; and Huxley Papers, ICL.HP.30.f.75.
101. Tylor, *Primitive Culture*, vol. 1, p. 6.
102. Ibid. Tylor did not state where he borrowed this proverb from, but it is allegedly a common expression in Italian.
103. Darwin, *Descent of Man*, vol. 1, p. 225.
104. Tylor, *Primitive Culture*, vol. 1, p. 6.
105. Ibid.
106. C. Holdsworth, 'Tylor, Edward Burnett (1832–1917), Anthropologist', *ODNB*.
107. For more on Tylor's comparative method, see Stocking, *Delimiting Anthropology*, p. 109.
108. For more information, see the Pitt-Rivers catalogue for 'Bushman Display from the Crystal Palace Exhibition', Pitt Rivers Coll., PRM 1998.2011.9; and 'Zulu Display from the Crystal Palace Exhibition', Pitt Rivers Coll., PRM 1998.2010.3.
109. Tylor, *Primitive Culture*, vol. 1, p. 7.
110. Ibid., p. 7.
111. Stocking, *Delimiting Anthropology*, p. 109.
112. Tylor, *Primitive Culture*, vol. 1., p. 8.
113. Ibid.
114. Ibid., p. 9.
115. Prichard, *Researches*, pp. 243–4; Lawrence, *Natural History of Man*, p. 109.
116. Tylor, *Primitive Culture*, vol. 1, p 11.
117. Ibid., p. 17.
118. Ibid., p. 16.
119. Ibid., p. 71.
120. Ibid., pp. 72–3.
121. Urry, *Before Social Anthropology*, p. 17.
122. C. Darwin, *The Expression of Emotions in Man and Animals* (London: John Murray, 1872). See also Radick, *The Simian Tongue*, pp. 31–7, 58–65, 386–94; G. Radick, 'Darwin's Puzzling Expression', *Comptes Rendus Biologies*, 333 (2010), pp. 181–7.
123. Browne, *Charles Darwin*, p. 11. For more on Darwin's use of correspondence networks, see J. Garber, 'Darwin's Correspondents in the Pacific: Through the Looking Glass to the Antipodes', in R. Macleod and P. Rehbock (eds), *Darwin's Laboratory: Evolutionary Theory and Natural History in the Pacific* (Honolulu, HI: University of Hawaii Press, 1994), pp. 169–211.
124. Prichard et al., 'Queries Respecting the Human Race'.
125. Darwin, *Expression of Emotions*, p. 15.
126. Ibid.
127. Hodgkin and Cull, 'A Manual of Ethnological Inquiry', p. 201.
128. C. Darwin to F. Müller, Darwin Correspondence, Letter number 5410, 22 February 1867, at http://www.darwinproject.ac.uk/entry-5410 [accessed 20 August 2012].
129. Ibid. These instructions were also reproduced in Darwin, *Expression of Emotions*, pp. 15–16. Darwin also sent the same instructions to other correspondents. For example, see: C. Darwin to A. R. Wallace, Darwin Correspondence, Letter number 5420, 26 February 1867, at http://www.darwinproject.ac.uk/entry-5420 [accessed 20 August

2012]; C. Darwin to J. Weale, Darwin Correspondence, Letter number 5422, 27 February 1867, at http://www.darwinproject.ac.uk/entry-5422 [accessed 20 August 2012]; C. Darwin to R. Swinhoe, Darwin Correspondence, Letter number 5423f, 27 February 1867, at http://www.darwinproject.ac.uk/entry-5423f [accessed 20 August 2012]; C. Darwin to J. Haast, Darwin Correspondence, Letter number 5423, 27 February 1867, at http://www.darwinproject.ac.uk/entry-5423 [accessed 20 August 2012].

130. Darwin to Müller, Darwin Correspondence, 22 February 1867.
131. Ibid.
132. J. Weale to C. Darwin, Darwin Correspondence, Letter number 5722, 10 December 1867, at http://www.darwinproject.ac.uk/entry-5722 [accessed 20 August 2012].
133. Hodgkin and Cull, 'A Manual of Ethnological Inquiry'.
134. R. M. Clements et al., 'Report of the Arctic Committee of the Anthropological Institute', *Journal of the Anthropological Institute of Great Britain and Ireland*, 2 (1873), pp. 291–306, on pp. 291–3.
135. J. B. Davis et al., 'Questions for Explorers (With Special Reference to Arctic Exploration)', *Journal of the Anthropological Institute of Great Britain and Ireland*, 2 (1873), pp. 296–306, on p. 296.
136. Ibid.
137. Ibid., p. 297.
138. Ibid.
139. See Huxley Papers, ICL.HP.30.f.75.
140. For more on British imperial networks, see: C. Hall (ed.), *Cultures of Empire: Colonizers in Britain and the Empire in the Nineteenth and Twentieth Centuries* (Manchester: Manchester University Press, 2000); Hall, McClelland and Rendall, *Defining the Victorian Nation*; C. Hall, *Civilising Subjects: Colony and Metropole in the English Imagination 1830–1867* (Chicago, IL: University of Chicago Press, 2002); and Laidlaw, *Colonial Connections 1815–1845*.
141. Davis et al., 'Questions for Explorers', p. 299.
142. Ibid., p. 304.
143. Ibid., p. 296.
144. See the second edition of Tylor, *Primitive Culture*, vol. 1, p. vii.
145. Ibid., pp. vii–viii.
146. Stocking, *Delimiting Anthropology*, p. 105.
147. Tylor, *Anthropology*; and Stocking, *Delimiting Anthropology*, p. 3.
148. E. B. Tylor, 'On a Method of Investigating the Development of Institutions; Applied to Laws of Marriage and Descent', *Journal of the Anthropological Institute*, 18 (1888), pp. 245–72.
149. E. Clodd, *The Birth and Growth of Myth, and its Survival in Folk-lore, Legend and Dogman* (London: Sunday Lecture Society, 1875); E. Clodd, *The Childhood of Religions: Embracing a Simple Account of the Birth and Growth of Myths and Legends* (London: Henry S. King and Co., 1876); J. Frazer, *Totemism* (Edinburgh: University of Edinburgh Press, 1887); and J. Frazer, *The Golden Bough: A Study of Magic and Religion* (London: Macmillan, 1890).

## Conclusion

1. A. C. Haddon, 'Review: Ethnology of New Guinea', *Geographical Journal*, 34 (1909), pp. 333–4, on p. 333.

2. For more on the Torres Straits Expedition, see A. Herle and S. Rouse (eds), *Cambridge and the Torres Straits: Centenary Essays on the 1898 Expedition* (Cambridge: Cambridge University Press, 1998). For discussions on sound recording, see Radick, *The Simian Tongue*, pp. 165–6. For discussions on cinematograph, see Grimshaw, *The Ethnographer's Eye*, pp. 20–5. For discussions on anthropology and photography, see E. Edwards (ed.), *Anthropology and Photography 1860–1920* (New Haven, CT: Yale University Press, 1992). Shaffer shows how anthropologists at the turn of the twentieth century – such as Haddon – transferred laboratory-based observational practices into the field; see Schaffer, *From Physics to Anthropology and Back Again*.
3. Stocking, *Victorian Anthropology*, p. 320. For more on the Torres Straits Expedition, see Herle and Rouse, *Cambridge and the Torres Straits*; and H. Kuklick, 'The Colour Blue: From Research in the Torres Straits to an Ecology of Human Behaviour', in R. MacLeod and P. Rehbock (eds), *Darwin's Laboratory: Evolutionary and Natural History in the Pacific* (Honolulu, HI: University of Hawaii Press, 1994), pp. 339–70.
4. Quoted in S. Rouse, 'Expedition and institution: A. C. Haddon and Anthropology at Cambridge', in A. Herle and S. Rouse (eds), *Cambridge and the Torres Straits: Centenary Essays on the 1898 Expedition* (Cambridge: Cambridge University Press, 1998), pp. 50–76, on p. 61.
5. H. J. Fleure, 'Alfred Cort Haddon 1855–1940', *Obituary Notices of Fellows of Fellows of the Royal Society*, 3 (1941), pp. 448–65, on pp. 454–6. See also Kuklick, 'The British Tradition', p. 63.
6. For more on Haddon's disciplinary map of anthropology, see Radick, *The Simian Tongue*, p. 162; and A. C. Haddon, *The Study of Man* (London: Bliss, Sans and Co., 1898), pp. xv–xx.
7. Quoted in A. H. Quiggin, *Haddon, the Head Hunter* (Cambridge: Cambridge University Press, 1942), p. 77. For more Haddon's field-centric approach, see Kuklick, 'After Ishmael', p. 47; and Kuklick, 'The British Tradition', p. 63.
8. Haddon, 'Review: Ethnology of New Guinea', p. 333.
9. Kuklick, *The Savage Within*, pp. 27–8.
10. Ibid., p. 27.
11. For a general overview of some of the changes occurring in tropical medicine during the second half of the nineteenth century, see D. Headrick, *The Tools of Empire: Technology and European Imperialism in the Nineteenth Century* (Oxford: Oxford University Press, 1981), pp. 66–75.
12. See Malinowski's entry from 2 September 1915 in B. Malinowski, *A Diary in the Strict Sense of the Term* (Palo Alto, CA: Stanford University Press, 1989), p. 84.
13. Kohler, *Landscapes and Labscapes*, pp. 82–5.
14. For more on Spencer and Gillen, see H. Kuklick, 'Humanity in the Crysalis Stage: Indigenous Australians in the Anthropological Imagination, 1899–1926', *British Journal for the History of Science*, 39 (2006), pp. 535–68.
15. For more, see Kuklick, 'Personal Equations', p. 9.
16. For more on Malinowki's anthropological programme, see Kuper, *Anthropology Anthropologists*, pp. 1–35; and Kuklick, 'Personal Equations'. See also Stocking, *Victorian Anthropology*, pp. 262–3.
17. Gupta and Ferguson, 'Discipline and Practice', p. 1; Angrisino and Crane, *Field Projects in Anthropology*, p. v; and Urry, 'A History of Field Methods', p. 27.

18. For more information of anthropology and its links to colonial development in the early twentieth century, consult Kuklick, *The Savage Within*, pp. 41–3; and Kuklick, 'The British Tradition', pp. 60–1.
19. For more, see Kuklick, 'Personal Equations', pp. 7–8. For more on transformations in sea and rail transport, see Headrick, *Tools of Empire*, pp. 27–33, 150–7, 192–201.
20. Headrick, *Tools of Empire*, pp. 150–7.
21. Erikson, 'What Makes School Ethnography "Ethnographic"?', p. 53.
22. Ellen, 'Introduction', p. 10.
23. Kuklick, 'Introduction', p. 1.
24. For more, see: Bleichmar, 'Training the Naturalist's Eye', p. 168; Daston and Galison, *Objectivity*, p. 19; and L. Daston and E. Lunbeck (eds), *Histories of Scientific Observation* (Chicago, IL: University of Chicago Press, 2011).
25. A. Cunningham, *The Anatomist Anatomis'd: An Experimental Discipline in Enlightenment Europe* (Farnham: Ashgate Publishing, 2010); and Bates, *The Anatomy of Robert Knox*.
26. R. Richardson, *The Making of Mr. Gray's Anatomy: Bodies, Books, Fortune, Fame* (Oxford: Oxford University Press, 2008).
27. Secord, *Artisan Naturalists*, pp. 135–206; and Daston and Galison, *Objectivity*.
28. Prichard devotes and entire volume in the third edition of his *Researches* to European ethnography. For other examples of nineteenth-century European ethnographic studies, see: Prichard, *The Eastern Origin of Celtic Nations*; Latham, *The Ethnology of Europe*; Latham, *Descriptive* Ethnology, vol. 1, pp. 421–5; and Knox, *Races of Man*, pp. 131–7.
29. Kaiser discusses at length the different intellectual trends that have emerged in PhD theses within the history of science over the past three decades. For further readings, see D. Kaiser, 'Training and the Generalist's Vision in the History of Science, *Isis*, 96 (2005), pp. 244–51. In particular, see footnote 12, which discusses other examples of recent secondary literature on disciplinary development and practice.
30. J. Morell and A. Thackray, *Gentlemen of Science: Early Years of the British Association for the Advancement of Science* (Oxford: Oxford University Press, 1981); G. Gooday, *The Morals of Measurement: Accuracy, Irony, and Trust in Late Victorian Electrical Practice* (Cambridge: Cambridge University Press, 2004); and R. O'Connor, *The Earth on Show: Fossils and the Poetics of Popular Science, 1802–1856* (Chicago, IL: University of Chicago Press, 2007).
31. Anon., 'Robert Gordon Latham, M.D., F.R.S.', p. 93.
32. Hall, *Civilising Subjects*; and Laidlaw, *Colonial Connections 1815–1845*.
33. Secord, 'Knowledge in Transit'.
34. For more, see Grasseni, *Skilled Visions*.
35. Grimshaw, *The Ethnographer's Eye*.
36. Daston and Lunbeck (eds), *Histories of Scientific Observation*.
37. See Sivasundaram, *Nature and the Godly Empire*. For more on methodological approaches to studying 'science and the global', see S. Sivasundaram, 'Sciences and the Global: On Methods, Questions and Theory', *Isis*, 101 (2010), pp. 146–58.
38. Qureshi, *Peoples on Parade*.
39. Topham, 'Scientific Publishing and the Reading of Science'; and Secord, 'Knowledge in Transit'.
40. For more on the French tradition, see Parkin, 'The French-Speaking Countries'; Sibeud, 'The Metamorphosis of Ethnology in France'; and Rabinow, *French Modern*.
41. The most obvious example is the Baudin expedition from 1800, where the Société des Observateurs de l'Homme sent naturalist François Péron to the South Pacific specifi-

cally to collect data on indigenous peoples from that region, and upon his return to publish his reports. See: Péron, *Voyage de Découverte aux Terres Australes*; de Gérando, *Considérations sur les Diverses Méthodes a Suivre*; de Gérando, *Considerations on the Diverse Methods to Follow*; J. Copans and J. Jamin (eds), *Aux Origines de l'Anthropologie Française: Les Mémoires de la Société des Observateurs de l'Homme en l'An VIII* (Paris: Sycomore, 1978); and Chappey, *La Société des Observateurs de l'Homme*.

42. For example, see F. Galton, 'Address to the Department of Anthropology', *Report of the British Association for the Advancement of Science*, 47 (1877), pp. 94–100.
43. For example, see: Stocking, 'On the Limits of "Presentism" and "Historicism"', pp. 3–6; Ellen, 'Introduction', pp. 1–3; Stocking, *Victorian Anthropology*, p. 285; Gupta and Ferguson, 'Discipline and Practice', p. 2; and Kuklick, 'Introduction', pp. 1–5.

# WORKS CITED

## Manuscript Sources

ASL Minutes: Minutes from the Proceedings of the Anthropological Society of London Meetings 1863–71, Anthropological Society of London Archive, Royal Anthropological Institute, London, UK.

Darwin Correspondence: http://www.darwinproject.ac.uk/, The Darwin Correspondence Project, University Library, University of Cambridge, UK.

ESL Minutes: Minutes from the Proceedings of the Ethnological Society of London Meetings 1844–5, Ethnological Society of London Archive, Royal Anthropological Institute, London, UK.

Huxley Papers: Thomas Henry Huxley Papers, College Archives, University College London, UK.

Notebook C: C. Darwin, 'Transmutation of Species (1838.02–1838.07)', Notebook C, Charles Darwin Archive, University Library, University of Cambridge, UK.

Pitt River Coll.: Pitt Rivers Photographic Collection, Pitt Rivers Museum, University of Oxford, UK.

PRO London: National Archives, Public Records Office, London, UK.

Tylor Notebooks: Pitt Rivers Manuscript Collection, Pitt Rivers Museum, University of Oxford, UK.

## Printed Sources

Angrisino, M., and J. Crane, *Field Projects in Anthropology: A Student Handbook* (Morristown, NJ: General Learning Press, 1974).

Anon., 'Review – Lectures on Physiology, &c.', *Imperial Magazine*, 4 (1822), pp. 668–74.

—, *An Account of the Life of William Burke, who was Executed on Wednesday the 28th Day of January, 1829, for Murder: With the Confession He Made while Under Sentence of Death* (Dunfermline: John Miller, 1829).

—, 'Researches into the Physical History of Mankind', *Athenaeum*, 776 (1842), pp. 797–8.

—, 'Regulations', *Journal of the Ethnological Society of London*, 1 (1848), pp. 3–14.

—, 'Art II. – The Natural History of the Varieties of Man', *Prospective Review*, 24 (1850), pp. 449–58.

—, 'Ethnology', *Literary Gazette*, 1765 (1850), pp. 855–7.

—, 'The Natural History of the Varieties of Man', *Examiner*, 2230 (1850), pp. 686–7.

—, 'The Varieties of Man', *Bentley's Miscellany*, 28 (1850), pp. 393–5.

—, 'The Natural History of the Varieties of Man', *Athenaeum*, 1226 (1851), pp. 447–8.

—, 'The Races of Mankind', *Fraser's Magazine for Town and Country*, 44 (1851), pp. 654–65.

—, 'Great Artists and Great Anatomists', *Athenaeum*, 1297 (1852), pp. 935–7.

—, 'The Crystal Palace at Sydenham', *Daily News*, Friday, 9 June 1854.

—, 'Anthropology at the British Association', *Anthropological Review*, 1 (1863), pp. 379–464.

—, 'Back Matters', *Transactions of the Ethnological Society*, 2 (1863), pp. 1–10 (back pages).

—, 'Robert Gordon Latham, M.D., F.R.S.', in L. Reeve (ed.), *Portraits of Men of Eminence in Literature, Science and Art, with Biographical Memoirs*, 6 vols (London: Lovell Reeve & Co., 1863), vol. 1, pp. 93–7.

—, 'Anthropology at the British Association, A.D. 1864', *Anthropological Review*, 2 (1864), pp. 294–335.

—, 'On the Prospects of Anthropological Science at the British Association of 1865', *Anthropological Review*, 3 (1865), pp. 224–9.

—, 'Back Matters', *Transactions of the Ethnological Society*, 5 (1867), pp. 1–12 (back pages).

—, 'Front Matters', *Journal of the Ethnological Society of London*, 1 (1869), pp. vi–xxviii.

—, 'A Sketch of the Life and Writings of Robert Knox, the Anatomist', *Examiner*, 3279 (1870), pp. 773–4.

—, 'Death of the Best Man in England', *Anthropological Review*, 8 (1870), p. 97.

Appel, T., *The Cuvier-Geoffroy Debate: French Biology in the Decades before Darwin* (Oxford: Oxford University Press, 1987).

Augstein, H. F., *James Cowles Prichard's Anthropology: Remaking the Science of Man in Early Nineteenth-Century Britain* (Amsterdam: Rodopi B.V., 1999).

—, 'Prichard, James Cowles (1786–1848), Physician and Ethnologist', *Oxford Dictionary of National Biography*.

Baigent, E., 'King, Richard (1810/11–1876), Artic Traveller and Ethnologist', *Oxford Dictionary of National Biography*.

Barfield, T. (ed.), *The Dictionary of Anthropology* (Oxford: Blackwell Publishing, 1997).

Barth, F., 'Britain and the Commonwealth', in F. Barth, A. Gingrich, R. Parkin and S. Silverman, *One Discipline Four Ways: British, German, French and American Anthropology* (Chicago, IL: University of Chicago Press, 2005), pp. 3–57.

Bartholomew, M., 'Huxley's Defence of Darwin', *Annals of Science*, 32 (1975), pp. 525–35.

Barton, R., 'Huxley, Lubbock, and Half a Dozen Others', *Isis*, 89 (1998), pp. 410–44.

Bates, A. W., *The Anatomy of Robert Knox: Murder, Mad Science and Medical Regulation in Nineteenth-Century Edinburgh* (Eastbourne: Sussex Academic Press, 2010).

Bibby, C., *Scientist Extraordinary: The Life and Scientific Work of Thomas Henry Huxley 1825–1895* (Oxford: Pergamon, 1972).

Bindman, D., *Ape to Apollo: Aesthetics and the Idea of Race in the 18th Century* (Ithaca, NY: Cornell University Press, 2002).

Blainey, G., *A Land Half Won* (London: Macmillan Press, 1980).

Blackman, H. J., 'Knox, Robert (1793–1862)', in B. Lightman (ed.), *Dictionary of Nineteenth-Century British Scientists*, 4 vols (London: Thoemmes Continuum, 2004), vol. 3, pp. 1160–2.

Blake, C. C., 'The Life of Robert Knox, the Anatomist. A Sketch of his Life and Writings by Henry Lonsdale', *Journal of Anthropology*, 1 (1871), pp. 332–8.

Bleichmar, D., 'Training the Naturalist's Eye in the Eighteenth Century: Perfect Global Visions and Local Blind Spots', in C. Grasseni (ed.), *Skilled Visions: Between Apprenticeship and Standards* (Oxford: Berghahn, 2007), pp. 166–90.

Blumenbach, J. F., *A Short System of Comparative Anatomy*, trans. W. Lawrence (London: Longman and Co., 1807).

—, *The Anthropological Treatises of Johann Friedrich Blumenbach, Late Professor at Göttingen and Court Physician to the King of Great Britain*, trans. T. Bendyshe (London: Longman, Green, Longman, Roberts, & Green, 1865).

Bougainville, L. A., *Voyage Autour du Monde: Par le Frégate du Roi la Boudeuse, et la Flûte l'Étoile* (Paris: Saillant and Nyon, 1771).

Broca, P., *On the Phenomena of Hybridity in the Genus Homo*, trans. C. C. Blake (London: Longman, Green Longman and Roberts, 1864).

Brantlinger, P., 'Victorians and Africans: The Genealogy of the Myth of the Dark Continent', *Critical Inquiry*, 12 (1985), pp. 166–203.

Bravo, M. T., 'Ethnological Encounters', in N. Jardine, J. A. Secord and E. C. Spary (eds), *Cultures of Natural History* (Cambridge: Cambridge University Press, 1996), pp. 338–57.

Brock, W. H., 'Brodie, Sir. Benjamin Collins, First Baronet (1783–1862), Physiologist and Surgeon', *Oxford Dictionary of National Biography*.

Brookes, H., and E. W. Brabrook, 'Anthropological Society of London', *Athenaeum*, 2132 (1868), p. 302.

Browne, J., *Charles Darwin: The Power of Place* (New York, NY: Alfred A. Knopf, 2002).

Burkhardt, F., D. Porter, J. Harvey and J. Topham, *The Correspondence of Charles Darwin*, 18 vols (Cambridge: Cambridge University Press, 1985– ).

Burrow, J. W., 'Evolution and Anthropology in the 1860's: The Anthropological Society of London, 1863–71', *Victorian Studies*, 7 (1963), pp. 137–49.

Calder, A., J. Lamb and B. Orr (eds), *Voyages and Beaches: Pacific Encounters, 1769–1840* (Honolulu, HI: University of Hawaii Press, 1999).

Cantor, G., *Quakers, Jews and Science: Religious Responses to Modernity and the Sciences in Britain 1650–1900* (Oxford: Oxford University Press, 2005).

—, *Religion and the Great Exhibition of 1851* (Oxford: Oxford University Press, 2012).

Cantor, G., and S. Shuttlesworth, 'Introduction', in G. Cantor and S. Shuttleworth (eds), *Science Serialized: Representation of the Sciences in Nineteenth-Century Periodicals* (Cambridge, MA: MIT Press, 2004), pp. 1–15.

Carpenter, W. B., 'Art V.-1 Researches into the Physical History of Mankind', *Edinburgh Review*, 88 (1848), pp. 429–87.

Caudill, E., 'The Bishop-Eaters: The Publicity Campaign for Darwin and on the Origin of Species', *Journal of the History of Ideas*, 55 (1994), pp. 441–60.

Cavell, J., *Tracing the Connected Narrative: Artic Exploration in British Print Culture, 1818–1860* (Toronto: University of Toronto Press, 2008).

Chappey, J. L., *La Société des Observateurs de l'Homme (1799–1804)* (Paris: Société des Études Robespierristes, 2002).

Clarke, H., 'Anthropological Society of London', *Athenaeum*, 2129 (1868), p. 210.

—, 'Anthropological Society of London', *Athenaeum*, 2131 (1868), pp. 271–2.

—, 'Anthropological Society of London', *Athenaeum*, 2143 (1868), pp. 681–2.

Clarke, H., and H. Brookes, 'Anthropological Society of London', *Athenaeum*, 2134 (1868), pp. 368–9.

Clarke, R., 'Sketches of the Colony of Sierra Leone and its Inhabitants', *Transactions of the Ethnological Society of London*, 2 (1863), pp. 320–63.

Clements, R. M., et al., 'Report of the Arctic Committee of the Anthropological Institute', *Journal of the Anthropological Institute of Great Britain and Ireland*, 2 (1873), pp. 291–306.

Clifford, J., and G. E. Marcus (eds), *Writing Culture: The Poetics and Politics of Ethnography* (Berkeley, CA: University of California Press, 1986).

Clodd, E., *The Birth and Growth of Myth, and its Survival in Folk-lore, Legend and Dogman* (London: Sunday Lecture Society, 1875).

—, *The Childhood of Religions: Embracing a Simple Account of the Birth and Growth of Myths and Legends* (London: Henry S. King and Co., 1876).

Coady, C. A. J., *Testimony: A Philosophical Study* (Oxford: Oxford University Press, 1992).

Collingwood, E. S., 'Death of Dr. James Hunt: Science or Sun-Stroke', *British Medical Journal*, 2 (1869), p. 355.

Conolly, J., *The Ethnological Exhibitions of London* (London: John Churchill, 1855).

Copans, J., and J. Jamin (eds), *Aux Origines de L'Anthropologie Française: Les Mémoires de la Société des Observateurs de l'Homme en L'An VIII* (Paris: Sycomore, 1978).

Crawfurd, J., 'On the Malayan and Polynesian Languages and Races', *Journal of the Ethnological Society of London*, 1 (1848), pp. 330–74.

—, *A Grammar and Dictionary of the Malay Language, with a Preliminary Dissertation*, 2 vols (London: Smith, Elder and Co., 1852).

Cull, R., 'Remarks on the Nature, Objects, and Evidences of Ethnological Science, an Address Read at a Conversazione on June 4, 1851', *Journal of the Ethnological Society of London*, 2 (1854), pp. 103–11.

Cunningham, A., *The Anatomist Anatomis'd: An Experimental Discipline in Enlightenment Europe* (Farnham: Ashgate Publishing, 2010).

Curtin, P. D., *The Image of Africa: British Ideas and Action, 1780–1850* (Madison, WI: University of Wisconsin Press, 1964).

Cuvier, G., 'Note Instructive sur les Recherches à Faire Relativement aux Différences Anatomiques des Diverses Races d'Hommes', in J. Copans and J. Jamin (eds), *Aux Origines de L'Anthropologie Française: Les Mémoires de la Société des Observateurs de l'Homme en L'An VIII* (Paris: Sycomore, 1978), pp. 171–6.

Dampier, W., *A Collection of Voyages*, 4 vols (London: James and John Knapton, 1729).

Darnell, R., 'North American Traditions in Anthropology: The Historiographic Baseline', in H. Kuklick (ed.), *A New History of Anthropology* (Oxford: Blackwell Publishing, 2008), pp. 35–51.

Darwin, C., *The Origin of Species by Means of Natural Selection or the Preservation of Favoured Races in the Struggle for Life* (London: John Murray, 1859).

—, *The Descent of Man, and Selection in Relation to Sex*, 2 vols (London: John Murray, 1871).

—, *The Expression of Emotions in Man and Animals* (London: John Murray, 1872).

—, *Beagle Diary*, ed. R. D. Keynes (Cambridge: Cambridge University Press, 2001).

Daston, L., and P. Galison, *Objectivity* (London: Zone Books, 2007).

Daston, L., and E. Lunbeck (eds), *Histories of Scientific Observation* (Chicago, IL: University of Chicago Press, 2011).

Davis, J. B., et al., 'Questions for Explorers (With Special Reference to Arctic Exploration)', *Journal of the Anthropological Institute of Great Britain and Ireland*, 2 (1873), pp. 296–306.

Dawson, G., R. Noakes and J. Topham, 'Introduction', in G. Cantor et al., *Science in the Nineteenth-Century Periodical: Reading the Magazine of Nature* (Cambridge: Cambridge University Press, 2004), pp. 1–34.

Dean, D., 'Jameson, Robert (1774–1854), Geologist and Natural Historian', *Oxford Dictionary of National Biography*.

De Gérando, J. M., *Considérations sur les Diverses Méthodes a Suivre dans l'Observation des Peuples Sauvages* (Paris: Societé des Observateurs de l'Homme, 1800).

—, *Considerations on the Diverse Methods to Follow in the Observations of Savage Peoples*, trans. F. C. T. Moore (London: Routledge & Kegan Paul, 1969).

Desmond, A., *The Politics of Evolution: Morphology, Medicine, and Reform in Radical London* (Chicago, IL: University of Chicago Press, 1989).

—, *Huxley: The Devil's Disciple* (London: Michael Joseph, 1994).

—, *Huxley: Evolution's High Priest* (London: Michael Joseph, 1997).

—, 'Redefining the X Axis: "Professionals", "Amateurs" and the Making of Mid-Victorian Biology – A Progress Report', *Journal of the History of Biology*, 34 (2001), pp. 3–50.

Desmond, A., and J. Moore, *Darwin* (London: Michael Joseph, 1991).

—, *Darwin's Sacred Cause: Race, Slavery and the Quest for Human Origins* (London: Penguin Books, 2009).

Dieffenbach, E., 'The Study of Ethnology', *Journal of the Ethnological Society of London*, 1 (1848), pp. 15–26.

Digby, S., and S. Simmonds (eds), *The Royal Asiatic Society: Its History and Treasures* (Leiden: Brill, 1979).

DiGregorio, M., *T. H. Huxley's Place in Natural Science* (New Haven, CT: Yale University Press, 1984).

Dixon, M., and G. Radick, *Darwin in Ilkley* (Stroud: History Press, 2009).

Edwards, E., 'Photographic Types: The Pursuit of Method', *Visual Anthropology*, 3 (1990), pp. 235–58.

— (ed.), *Anthropology and Photography 1860–1920* (New Haven, CT: Yale University Press, 1992).

—, *Raw Histories: Photographs, Anthropology and Museums* (Oxford: Berg, 2001).

Edwards, E., and C. Morton, 'Introduction', in E. Edwards and C. Morton (eds), *Photography, Anthropology and History: Expanding the Frame* (Farnham: Ashgate, 2009), pp. 1–24.

Elwick, J., *Styles of Reasoning in the British Life Sciences: Shared Assumptions, 1820–1858* (London: Pickering & Chatto, 2007).

Endersby, J., *Imperial Nature: Joseph Hooker and the Practice of Victorian Science* (Chicago, IL: University of Chicago Press, 2008).

Efron, J., *Defenders of Race: Jewish Doctors and Race Science in Fin-de-Siècle Europe* (New Haven, CT: Yale University Press, 1994).

Ellen, R. F., 'Introduction', in R. F. Ellen (ed.), *Ethnographic Research: A Guide to General Conduct* (London: Academic Press, 1984), pp. 1–12.

Erikson, F., 'What Makes School Ethnography "Ethnographic"?', *Anthropology & Education Quarterly*, 14 (1984), pp. 51–66.

Fee, E., 'The Sexual Politics of Victorian Social Anthropology', *Feminist Studies*, 1 (1973), pp. 23–39.

Fichman, M., 'Biology and Politics: Defining the Boundaries', in B. Lightman (ed.), *Victorian Science in Context* (Chicago, IL: University of Chicago Press, 1997), pp. 94–118.

Findlen, P., *Possessing Nature: Museums, Collecting and Scientific Culture in Early Modern Italy* (Berkeley, CA: University of California Press, 1994).

Fitzroy, R., and P. P. King, *Narrative of the Surveying Voyages of His Majesty's Ships Adventure and Beagle, between the Years 1826–1836, Describing their Examination of the Southern Shores of South America, and the Beagle's Circumnavigation of the Globe*, 3 vols (London: Henry Colburn, 1839).

Fleure, H. J., 'Alfred Cort Haddon 1855–1940', *Obituary Notices of Fellows of the Royal Society*, 3 (1941), pp. 448–65.

Forbes, E., and R. G. Latham, *The Natural History Department of the Crystal Palace Described* (London: Crystal Palace Library, 1854).

Forgan, S., 'The Architecture of Display: Museums, Universities and Objects in Nineteenth-Century Britain', *History of Science*, 32 (1994), pp. 139–62.

Forster, J. R., *The Resolution Journal of Johann Reinhold Forster*, ed. M. Hoare, 4 vols (London: Hakluyt Society, 1982).

Frazer, J., *Totemism* (Edinburgh: University of Edinburgh Press, 1887).

—, *The Golden Bough: A Study of Magic and Religion* (London: Macmillan, 1890).

Galton, F., 'Address to the Department of Anthropology', *Report of the British Association for the Advancement of Science*, 47 (1877), pp. 94–100.

Garber, J., 'Darwin's Correspondents in the Pacific: Through the Looking Glass to the Antipodes', in R. Macleod and P. Rehbock (eds), *Darwin's Laboratory: Evolutionary Theory and Natural History in the Pacific* (Honolulu, HI: University of Hawaii Press, 1994), pp. 169–211.

Gathercole, P., 'Cambridge and the Torres Straits, 1888–1920', *Cambridge Anthropology*, 3 (1977), pp. 22–31.

George, W. H., 'Alfred Tylor (1824–1884), Geologist and Brass-Founder', *Oxford Dictionary of National Biography*.

Gooday, G., '"A Fungoid Assemblage of Buildings": Diversity and Adversity in the Development of College Architecture and Scientific Education in Nineteenth-Century South Kensington', *History of Universities*, 13 (1994), pp. 153–92.

—, 'Constructing South Kensington: The Buildings and Politics of T. H. Huxley's Working Environments', *British Journal for the History of Science*, 29 (1996), pp. 435–68.

—, *The Morals of Measurement: Accuracy, Irony, and Trust in Late Victorian Electrical Practice* (Cambridge: Cambridge University Press, 2004).

Goodfield-Toulmin, J., 'Some Aspects of English Physiology: 1780–1840', *Journal of the History of Biology*, 2 (1969), pp. 283–320.

Grasseni, C. (ed.), *Skilled Visions: Between Apprenticeship and Standards* (Oxford: Berghahn, 2007).

Grimshaw, A., *The Ethnographer's Eye: Ways of Seeing in Modern Anthropology* (Cambridge: Cambridge University Press, 2001).

Gruber, H., *Darwin on Man: A Psychological Study of Scientific Creativity* (London: Wildwood House London, 1974).

Gupta, A., and J. Ferguson, 'Discipline and Practice: "The Field" as Site, Method, and Location in Anthropology', in A. Gupta and J. Ferguson (eds), *Anthropological Locations: Boundaries and Grounds of a Field Science* (Berkeley, CA: University of California Press, 1997), pp. 1–46.

Haddon, A. C., *The Study of Man* (London: Bliss, Sans and Co., 1898).

—, 'Review: Ethnology of New Guinea', *Geographical Journal*, 34 (1909), pp. 333–4.

Haddon, A. C., and A. H. Quiggin, 'Introduction', in A. C. Haddon and A. H. Quiggin, *History of Anthropology* (London: G. P. Putman's Sons, 1910), pp. ix–xiv.

Hall, C. (ed.), *Cultures of Empire: Colonizers in Britain and the Empire in the Nineteenth and Twentieth Centuries* (Manchester: Manchester University Press, 2000).

—, *Civilising Subjects: Colony and Metropole in the English Imagination 1830–1867* (Chicago, IL: University of Chicago Press, 2002).

Hall, C., K. McClelland and J. Rendall, *Defining the Victorian Nation: Class, Race, Gender and the Reform Act of 1867* (Cambridge: Cambridge University Press, 2000).

Haller, J., 'Concepts of Race Inferiority in Nineteenth-Century Anthropology', *Journal of the History of Medicine*, 25 (1970), pp. 40–51.

Harding, C., and D. I. Heath, 'Anthropological Society of London', *Athenaeum*, 2133 (1868), p. 334.

Headrick, D., *The Tools of Empire: Technology and European Imperialism in the Nineteenth Century* (Oxford: Oxford University Press, 1981).

Heartfield, J., *The Aborigines' Protection Society: Humanitarian Imperialism in Australia, New Zealand, Fiji, Canada, South Africa, and the Congo, 1836–1909* (New York: Columbia University Press, 2011).

Herle, A., and S. Rouse (eds), *Cambridge and the Torres Straits: Centenary Essays on the 1898 Expedition* (Cambridge: Cambridge University Press, 1998).

Herschel, J. F. W. (ed.), *A Manual of Scientific Enquiry; Prepared for the Use of Officers in Her Majesty's Navy and Travellers in General* (London: John Murray, 1859).

Hodgkin, T., 'The Progress of Ethnology', *Journal of the Ethnological Society of London*, 1 (1848), pp. 27–45.

—, 'Obituary of Dr. Prichard', *Journal of the Ethnological Society of London*, 2 (1850), pp. 182–207.

Hodgkin, T., and R. Cull, 'A Manual of Ethnological Inquiry', *Journal of the Ethnological Society of London*, 3 (1854), pp. 193–208.

Holdsworth, C., 'Tylor, Edward Burnett (1832–1917), Anthropologist', *Oxford Dictionary of National Biography*.

Hunt, J., *Stammering and Stuttering, their Nature and Treatment* (London: Longman, Green Longman and Roberts, 1861).

—, 'Abstract of a Report of the Ethnological Papers Read at the Manchester Meeting of the British Association in September 1861', *Transactions of the Ethnological Society of London*, 2 (1863), pp. 1–3.

—, 'Introductory Address on the Study of Anthropology', *Anthropological Review*, 1 (1863), pp. 1–20.

—, *On the Negro's Place in Nature* (London: Trübner and Co., 1863).

—, 'Dedication to Dr. Paul Broca', in C. Vogt, *Lectures on Man: His Place in Creation, and in the History of the Earth*, ed. J. Hunt (London: Longman, Green, Longman and Roberts, 1864), pp. v–x.

—, 'The President's Address', *Journal of the Ethnological Society of London*, 2 (1864), pp. lxxx–xcv.

—, 'Address Delivered at the Third Anniversary Meeting of the Anthropological Society of London', *Journal of the Anthropological Society of London*, 4 (1866), pp. lix–lxxxi.

—, 'On the Application of the Principle of Natural Selection to Anthropology, in Reply to the Views Advocated by Some of Mr Darwin's Disciples', *Anthropological Review*, 4 (1866), pp. 320–40.

—, 'Anthropological News', *Anthropological Review*, 5 (1867), pp. 240–56.

—, 'On the Doctrine of Continuity Applied to Anthropology', *Anthropological Review*, 5 (1867), pp. 110–20.

—, 'On Physio-Anthropology, its Aim and Method', *Journal of the Anthropological Society of London*, 5 (1867), pp. ccix–cclxxi.

—, 'The President's Address', *Journal of the Anthropological Society of London*, 5 (1867), pp. xliv–lxxi.

—, 'Anthropological Society of London', *Athenaeum*, 2135 (1868), p. 402.

—, 'On the Origin of the Anthropological Review and its Connection with the Anthropological Society', *Anthropological Review*, 6 (1868), pp. 431–42.

Hunt, J., and J. F. Collingwood, 'Anthropological Society of London', *Athenaeum*, 2130 (1868), pp. 239–40.

Huxley, J. (ed.), *T. H. Huxley's Diary of the Voyage of H.M.S. Rattlesnake* (London: Chatto and Windus, 1935).

Huxley, L. (ed.), *Life and Letters of Thomas Henry Huxley*, 2 vols (London: Macmillan and Co., 1900).

Huxley, T. H., 'Time and Life: Mr. Darwin's "Origin of Species"', *Macmillan's Magazine*, 2 (1859), pp. 142–8.

—, *Evidence as to Man's Place in Nature* (London: Williams and Norgate, 1863).

—, 'Emancipation – Black and White', *Reader*, 5 (1865), pp. 561–2.

—, 'On the Methods and Results of Ethnology', *Fortnightly Review*, 1 (1865), pp. 257–77.

—, 'On the Ethnology and Archaeology of India: Opening Address of the President', *Journal of Ethnological Society of London*, 1 (1869), pp. 89–93.

Irvine, W., *Thomas Henry Huxley* (London: Longmans, 1960).

Isichei, E., *Victorian Quakers* (Oxford: Oxford University Press, 1970).

Jacyna, L. S., 'Immanence or Transcendence: Theories of Life and Organisation in Britain, 1790–1835', *Isis*, 74 (1983), pp. 311–29.

—, *Philosophical Whigs: Medicine, Science and Citizenship in Edinburgh, 1789–1848* (London: Routledge, 1994).

—, 'Lawrence, Sir. William, First Baronet (1783–1867), Surgeon', *Oxford Dictionary of National Biography*.

James, A., et al., *After Writing Culture: Epistemology and Praxis in Contemporary Anthropology* (London: Routledge, 1997).

Jensen, J. V., *Thomas Henry Huxley: Communicating for Science* (Newark, DE: University of Delaware Press, 1991).

Kaiser, D., 'Training and the Generalist's Vision in the History of Science', *Isis*, 96 (2005), pp. 244–51.

Kass, A. M., 'Hodgkin, Thomas (1798–1866), Physician and Social Reformer', *Oxford Dictionary of National Biography*.

Kass, A. M., and E. H. Kass, *Perfecting the World: The Life and Times of Dr. Thomas Hodgkin 1798–1866* (Boston, MA: Harcourt Brace Jovanovich, 1988).

Kenny, R., 'From the Curse of Ham to the Curse of Nature: The Influence of Natural Selection on the Debate on Human Unity before the Publication of the "Descent of Man"', *British Journal for the History of Science*, 40 (2007), pp. 367–88.

Keynes, R. D., 'Introduction', in C. Darwin, *Beagle Diary*, ed. R. D. Keynes (Cambridge: Cambridge University Press, 2001), pp. xi–xxiv.

King, R., 'On the Physical Characters of the Esquimaux', *Journal of the Ethnological Society of London*, 1 (1848), pp. 45–59.

—, 'On the Intellectual Character of the Esquimaux', *Journal of the Ethnological Society of London*, 1 (1848), pp. 127–53.

—, 'On the Industrial Arts of the Esquimaux', *Journal of the Ethnological Society of London*, 1 (1848), pp. 277–300.

—, 'Address to the Ethnological Society of London Delivered at the Anniversary, 25th May 1844', *Journal of the Ethnological Society of London*, 2 (1850), pp. 9–42.

Knox, R., *The Races of Man: A Fragment* (London: Henry Renshaw, 1850).

—, *Great Artists and Great Anatomists* (London: John Van Voorst, 1852).

—, *A Manual of Human Anatomy* (London: Henry Renshaw, 1853).

Kohler, R., *Landscapes and Labscapes: Exploring the Lab-Field Border in Biology* (Chicago, IL: University of Chicago Press, 2002).

Krupat, A., *Ethnocentrism: Ethnography, History, Literature* (Berkeley, CA: University of California Press, 1992).

Kuklick, H., 'The Sins of the Fathers: British Anthropology and African Colonial Administration', *Research in Sociology of Knowledge, Science and Art*, 1 (1978), pp. 93–119.

—, *Imperial Bureaucrat: Colonial Administrative Service in the Gold Coast 1920–1939* (Stanford, CA: Hoover Institution Press, 1979).

—, *The Savage Within: The Social History of British Anthropology, 1885–1945* (Cambridge: Cambridge University Press, 1991).

—, 'The Colour Blue: From Research in the Torres Straits to an Ecology of Human Behaviour', in R. MacLeod and P. Rehbock (eds), *Darwin's Laboratory: Evolutionary and Natural History in the Pacific* (Honolulu, HI: University of Hawaii Press, 1994), pp. 339–70.

—, 'After Ishmael: The Fieldwork Tradition and its Future', in A. Gupta and J. Ferguson (eds), *Anthropological Locations: Boundaries and Grounds of a Field Site* (Berkeley, CA: University of California Press, 1997), pp. 47–65.

—, 'Humanity in the Crysalis Stage: Indigenous Australians in the Anthropological Imagination, 1899–1926', *British Journal for the History of Science*, 39 (2006), pp. 535–68.

—, 'Introduction', in H. Kuklick (ed.), *A New History of Anthropology* (Oxford: Blackwell Publishing, 2008), pp. 1–16.

—, 'The British Tradition', in H. Kuklick (ed.), *A New History of Anthropology* (Oxford: Blackwell Publishing, 2008), pp. 52–78.

—, 'Personal Equations: Reflections on the History of Fieldwork, with Special Reference to Sociocultural Anthropology', *Isis*, 102 (2011), pp. 1–33.

Kuper, A., *Anthropology Anthropologists: The Modern British School*, 3rd edn (London: Routledge, 1993).

Laidlaw, Z., *Colonial Connections 1815–1845: Patronage, the Information Revolution and Colonial Government* (Manchester: Manchester University Press, 2005).

—, 'Heathens, Slaves and Aborigines: Thomas Hodgkin's Critique of Missions and Anti-slavery', *History Workshop Journal*, 64 (2007), pp. 134–61.

Lamprey, J. H., and H. Clark, 'Ethnological Notes and Queries', *Journal of the Ethnological Society of London*, 1 (1869), pp. 84–8.

Lang, A., 'Edward Burnett Tylor', in H. Balfour (ed.), *Anthropological Essays Presented to Edward Burnett Tylor in Honour of his 75th Birthday* (Oxford: Clarendon Press, 1907), pp. 1–17.

Latham, R. G., *An Elementary English Grammar, for the Use of Schools* (London: Taylor, Walton and Maberly, 1843).

—, 'On the Languages of the Oregon Territory', *Journal of the Ethnological Society of London*, 1 (1848), pp. 154–66.

—, 'On the Ethnography of Russian America', *Journal of the Ethnological Society of London*, 1 (1848), pp. 182–91.

—, *History and Etymology of the English Language* (London: Taylor, Walton and Maberly, 1849).

—, *The Natural History of the Varieties of Man* (London: John Van Voorst, 1850).

—, *A Hand-Book of the English Language: For the Use of Students of the Universities and Higher Classes of School* (London: Taylor, Walton and Maberly, 1851).

—, 'General and Special Apophthegms', *Edinburgh New Philosophical Journal*, 51 (1851), pp. 39–44.

—, *The Ethnology of Europe* (London: John Van Voorst, 1852).

—, *Descriptive Ethnology*, 2 vols (London: John Van Voorst, 1859).

—, *Elements of Comparative Philology* (London: Taylor, Walton and Maberly, 1862).

—, *A Dictionary of the English Language*, 2 vols (London: Longmans and Green, 1864–70).

—, *Essential Rules and Principles for the Study of English Grammar* (London: Longmans and Green, 1876).

—, 'Ethnology', in E. Forbes and R. G. Latham, *The Natural History Department of the Crystal Palace Described* (London: Crystal Palace Library, 1854), pp. 5–79.

—, 'The Antiquity of the Human Race', *Westminster Review*, 74 (1860), pp. 411–36.

Lawrence, W., *An Introduction to Comparative Anatomy and Physiology; Being the Two Introductory Lectures Delivered at the Royal College of Surgeons, on the 21st and 25th of March, 1816* (London: J. Callow Medical Booksellers, 1816).

—, *Lectures on Physiology, Zoology, and the Natural History of Man, Delivered at the Royal College of Surgeons* (Salem, NY: Foote and Brown, 1828).

Leopold, J., *Culture in Comparative and Evolutionary Perspective: E. B. Tylor and the Making of Primitive Culture* (Berlin: Dietrich Reimer Verlag, 1980).

Levine, P. J. A., *The Amateur and the Professional: Antiquarians, Historians and Archaeologists in Victorian England 1838–1886* (Cambridge: Cambridge University Press, 2003).

Liebersohn, H., 'Anthropology before Anthropology', in H. Kuklick (ed.), *A New History of Anthropology* (Oxford: Blackwell Publishing, 2008), pp. 17–31.

Lightman, B., *The Origins of Agnosticism: Victorian Unbelief and the Limits of Knowledge* (Baltimore, MD: John Hopkins University Press, 1987).

—, *Victorian Popularizers of Science: Designing Nature for New Audiences* (Chicago, IL: University of Chicago Press, 2007).

Linnaeus, C., *The Elements of Botany: Containing the History of Science*, trans. H. Rose (London: Cadell, 1775).

Livingstone, D., *Adam's Ancestors: Race, Religion and the Politics of Human Origins* (Baltimore, MD: Johns Hopkins University Press, 2008).

Long, G. (ed.), *Penny Cyclopaedia of the Society for the Diffusion of Useful Knowledge*, 12 vols (London: Charles Knight, 1833–43).

Lonsdale, H., *A Sketch of the Life and Writings of Robert Knox, the Anatomist* (London: Macmillan, 1870).

Lorimer, D., *Colour, Class and the Victorians: English Attitudes to the Negro in the Mid-Nineteenth-Century* (Leicester: Leicester University Press, 1978).

—, 'Hunt, James 1833–1869', in B. Lightman (ed.), *Dictionary of Nineteenth-Century British Scientists*, 4 vols (London: Thoemmes Continuum, 2004), vol. 2, pp. 1033–5.

Love, R., *Maritime Exploration in the Age of Discovery, 1415–1800* (London: Greenwood Publishing Group, 2006).

Lyons, S., *Thomas Henry Huxley: The Evolution of a Scientist* (Amherst, NY: Prometheus Books, 1999).

Macleod, R., 'The X Club a Social Network of Science in Late-Victorian England', *Notes and Records of the Royal Society of London*, 24 (1970), pp. 305–22.

Malcolm, C., 'Address to the Ethnological Society of London, Delivered at the Anniversary, 14th May, 1851', *Journal of the Ethnological Society of London*, 3 (1854), pp. 86–102.

Malinowski, B., *Argonauts of the Western Pacific: An Account of the Native Enterprise and Adventure in the Archipelagoes of Melanesian New Guinea*, 2nd edn (London: George Routledge & Sons, Ltd, 1932).

—, *A Diary in the Strict Sense of the Term* (Palo Alto, CA: Stanford University Press, 1989).

Man, E. H., 'On the Aboriginal Inhabitants of the Andaman Islands (Part I)', *Journal of the Anthropological Institute of Great Britain and Ireland*, 12 (1882), pp. 69–116.

—, 'On the Aboriginal Inhabitants of the Andaman Islands (Part II)', *Journal of the Anthropological Institute of Great Britain and Ireland*, 12 (1882), pp. 117–75.

—, 'On the Aboriginal Inhabitants of the Andaman Islands (Part III)', *Journal of the Anthropological Institute of Great Britain and Ireland*, 12 (1882), pp. 327–434.

Martineau, H., *How to Observe: Morals and Manners* (London: Charles Knight and Co., 1838).

Miller, D. P., *Visions of Empire: Voyages, Botany, and Representations of Nature* (Cambridge: Cambridge University Press, 1996).

Miller, T., *Colonizing Egypt* (Cambridge: Cambridge University Press, 1988).

Mitford, J., 'The Races of Men: A Fragment. By Robert Knox M.D.', *Gentleman's Magazine* (November 1850), pp. 528–9.

Morrell, J., 'The Legacy of William Smith: The Case of John Phillips in the 1820s', *Archives of Natural History*, 16 (1989), pp. 319–35.

—, *John Phillips and the Business of Victorian Science* (Aldershot: Ashgate, 2005).

Morell, J., and A. Thackray, *Gentlemen of Science: Early Years of the British Association for the Advancement of Science* (Oxford: Oxford University Press, 1981).

Nord, D. E., *Gypsies and the British Imagination, 1807–1930* (New York: Columbia University Press, 2008).

Obeyesekere, G., *The Apotheosis of Captain Cook: European Mythmaking in the Pacific* (Princeton, NJ: Princeton University Press, 1992).

O'Connor, R., *The Earth on Show: Fossils and the Poetics of Popular Science, 1802–1856* (Chicago, IL: University of Chicago Press, 2007).

Okely, J., *The Traveller-Gypsies* (Cambridge: Cambridge University Press, 1983).

Parkin, R., 'The French-Speaking Countries', in F. Barth, A. Gingrich, R. Parkin and S. Silverman, *One Discipline Four Ways: British, German, French and American Anthropology* (Chicago, IL: University of Chicago Press, 2005), pp. 157–253.

Paul, P., 'Darwin, Social Darwinism and Eugenics', in J. Hodge and G. Radick (eds), *The Cambridge Companion to Darwin* (Cambridge: Cambridge University Press, 2003), pp. 214–40.

Péron, F., *Voyage de Découverte aux Terres Australes* (Paris: Arthus Bertrand Libraire Éditeur, 1824).

Porter, R., *The Making of Geology: Earth Science in Britain 1660–1815* (Cambridge: Cambridge University Press, 1977).

Pouchet, G., *On the Plurality of the Human Race*, trans. H. Beaven (London: Longman, Green, Longman and Roberts, 1864).

Prichard, J. C., 'Of the Varieties of the Human Race', *Records of the Royal Medical Society of Edinburgh*, 58 (1807–8), pp. 87–134.

—, *Researches into the Physical History of Man* (London: John and Arthur Arch, 1813).

—, *Researches into the Physical History of Mankind*, 2nd edn, 2 vols (London: John and Arthur Arch, 1826).

—, *The Eastern Origin of Celtic Nations Proved by a Comparison to Sanskrit, Greek, Latin and Teutonic Languages* (London: John and Arthur Arch, 1831).

—, *A Treatise on Insanity* (London: Marchant, 1833).

—, *A Treatise on Insanity, and Other Disorders Affecting the Mind* (London: Sherwood, 1835).

—, *Researches into the Physical History of Mankind*, 3rd edn, 5 vols (London: Sherwood, Gilbert and Piper, 1836–47).

—, *Naturgeschichte des Menschengeschletchs*, trans. R. Wagner and J. G. F. Will, 5 vols (Leipzig: Verlag von Leopold Vos., 1840–8).

—, *On the Different Forms of Insanity, in Relation to Jurisprudence. Designed for the Use of Persons Concerned in Legal Questions Regarding Unsoundness of Mind* (London: H. Baillière, 1842).

—, *The Natural History of Man: Comprising Inquiries into the Modifying Influence of Physical and Moral Agencies on the Different Tribes of the Human Family*, 1st edn (London: H. Baillière, 1843).

—, *Histoire Naturelle de l'Homme*, trans. F. Roulin (Paris: J. Baillière, 1843).

—, 'Ethnology', in J. Herschel (ed.), *A Manual of Scientific Enquiry: Prepared for the Use of Officers in Her Majesty's Navy and Travellers in General* (London: John Murray, 1849), pp. 423–40.

—, 'Anniversary Address for 1848, to the Ethnological Society of London on the Recent Progress of Ethnology', *Journal of the Ethnological Society of London*, 2 (1850), pp. 119–49.

—, *Researches into the Physical History of Mankind*, 4th edn, 5 vols (London: Houlston, and Stoneman, 1851–8).

Prichard, J. C., et al., 'Queries Respecting the Human Race to be Addressed to Travellers and Others', in Anon. (ed.), *Report of the Eleventh Meeting of the British Association for the Advancement of Science* (London: John Murray, 1842), pp. 332–9.

Quiggin, A. H., *Haddon, the Head Hunter* (Cambridge: Cambridge University Press, 1942).

Qureshi, S., *Peoples on Parade: Exhibitions, Empire and Anthropology in Nineteenth-Century Britain* (Chicago, IL: University of Chicago Press, 2011).

—, 'Robert Gordon Latham, Displayed Peoples and the Natural History of Race, 1854–1866', *Historical Journal*, 54 (2011), pp. 143–66.

Rabinow, P., *French Modern: Norms and Forms of the Social Environment* (Cambridge, MA: MIT Press, 1989).

Radick, G., *The Simian Tongue: The Long Debate about Animal Language* (Chicago, IL: University of Chicago Press, 2007).

—, 'Darwin's Puzzling Expression', *Comptes Rendus Biologies*, 333 (2010), pp. 181–7.

—, 'Did Darwin Change his Mind about the Fuegians?', *Endeavour*, 34 (2010), pp. 50–4.

Rae, I., *Knox the Anatomist* (Edinburgh: Oliver and Boyd, 1964).

Richards, E., 'Huxley and Woman's Place in Science: The Woman Question and the Control of Victorian Anthropology', in J. Moore (ed.), *History, Humanity and Evolution* (Cambridge: Cambridge University Press, 1989), pp. 253–84.

—, 'The "Moral Anatomy" of Robert Knox: The Interplay between Biological and Social Thought in Victorian Scientific Naturalism', *Journal of the History of Biology*, 22 (1989), pp. 373–436.

Richardson, R., *Death Dissection and the Destitute* (London: Routledge & Kegan Paul, 1987).

—, *The Making of Mr. Gray's Anatomy: Bodies, Books, Fortune, Fame* (Oxford: Oxford University Press, 2008).

Ridler, A. M., 'Latham, Robert Gordon (1812–1888), Ethnologist and Philologist', *Oxford Dictionary of National Biography*.

Rodan, A. A., 'Looking at Anthropology from a Biological Point of View: Haddon's Metaphors on Anthropology', *History of the Human Sciences*, 5 (1992), pp. 21–32.

Rosner, L., *Medical Education in the Age of Improvement: Edinburgh Students and Apprentices, 1760–1826* (Edinburgh: Edinburgh University Press, 1991).

—, 'Monro, Alexander, Secundus (1733–1817), Anatomist', *Oxford Dictionary of National Biography*.

Rouse, S., 'Expedition and Institution: A. C. Haddon and Anthropology at Cambridge', in A. Herle and S. Rouse (eds), *Cambridge and the Torres Straits: Centenary Essays on the 1898 Expedition* (Cambridge: Cambridge University Press, 1998), pp. 50–76.

Rudwick, M., *The Great Devonian Controversy: The Shaping of Scientific Knowledge among Gentlemanly Specialists* (Chicago, IL: University of Chicago Press, 1985).

Russell, G., 'An "Entertainment of Oddities": Fashionable Sociability and the Pacific in the 1770s', in K. Wilson (ed.), *A New Imperial History: Culture, Identity and Modernity in Britain and the Empire 1660–1840* (Cambridge: Cambridge University Press, 2008), pp. 48–70.

Schaffer, S., *From Physics to Anthropology and Back Again* (Cambridge: Prickly Pear Press, 1994).

Schaffer, S., and J. Delbourgo et al., *The Brokered World: Go-Betweens and Global Intelligence 1770–1820* (Sagamore Beach, MA: Science History Publications, 2009).

Scherer, J., 'The Photographic Document: Photographs as Primary Data in Anthropological Enquiry', in E. Edwards (ed.), *Anthropology and Photography 1860–1920* (New Haven, CT: Yale University Press, 1992), pp. 32–41.

Secord, A., 'Corresponding Interests: Artisans and Gentlemen in Nineteenth-Century Natural History', *British Journal for the History of Science*, 27 (1994), pp. 383–408.

—, *Artisan Naturalists: Science as Popular Culture in Nineteenth-Century England* (PhD thesis, University of London, 2002).

Secord, J., 'Knowledge in Transit', *Isis*, 95 (2004), pp. 654–72.

Sera-Shriar, E., 'Ethnology in the Metropole: Robert Knox, Robert Gordon Latham and Local Sites of Observational Training', *Studies in History and Philosophy of Biological and Biomedical Sciences*, 42 (2011), pp. 486–96.

—, 'Observing Human Difference: James Hunt, Thomas Huxley and Competing Disciplinary Strategies in the 1860s', *Annals of Science*, forthcoming.

Shapin, S., *A Social History of Truth: Civility and Science in Seventeenth-Century England* (Chicago, IL: University of Chicago Press, 1994).

Sibeud, E., 'The Metamorphosis of Ethnology in France, 1839–1930', in H. Kuklick (ed.), *A New History of Anthropology* (Oxford: Blackwell Publishing, 2008), pp. 96–110.

Silverman, S., 'The United States', in F. Barth, A. Gingrich, R. Parkin and S. Silverman, *One Discipline Four Ways: British, German, French and American Anthropology* (Chicago, IL: University of Chicago Press, 2005), pp. 257–347.

Sivasundaram, S., *Nature and the Godly Empire: Science and Evangelical Mission in the Pacific 1795–1850* (Cambridge: Cambridge University Press, 2005).

—, 'Sciences and the Global: On Methods, Questions and Theory', *Isis*, 101 (2010), pp. 146–58.

Spencer, F., 'Hunt, James (1833–1869)', in F. Spencer (ed.), *History of Physical Anthropology*, 2 vols (London: Garland Publishing, Inc., 1997), vol. 1, pp. 506–8.

Stagl, J., *A History of Curiosity: The Theory of Travel 1550–1800* (Chur: Harwood Academic Publishers, 1995).

Stocking, G., 'French Anthropology in 1800', *Isis*, 55 (1964), pp. 134–50.

—, 'What's in a Name? The Origins of the Royal Anthropological Institute (1837–71)', *Man*, 6 (1971), pp. 369–90.

—, 'From Chronology to Ethnology: James Cowles Prichard and British Anthropology 1800–1850', in J. C. Prichard, *Researches into the Physical History of Man*, ed. G. Stocking (Chicago, IL: University of Chicago Press, 1973), pp. ix–cx.

—, 'Bibliography of James Cowles Prichard', in J. C. Prichard, *Researches into the Physical History of Man*, ed. G. Stocking (Chicago, IL: University of Chicago Press, 1973), pp. cxi–cxviii.

—, 'Guide to Prichard's References', in J. C. Prichard, *Researches into the Physical History of Man*, ed. G. Stocking (Chicago, IL: University of Chicago Press, 1973), pp. cxix–cxliv.

—, 'On the Limits of "Presentism" and "Historicism" in the Historiography of the Behavioural Sciences', in G. Stocking, *Race, Culture, and Evolution: Essays in the History of Anthropology* (London: Blackwell, 1982), pp. 1–12.

—, *Victorian Anthropology* (New York: Free Press, 1987).

—, 'The Ethnographer's Magic: Fieldwork in British Anthropology from Tylor to Malinowski', in G. Stocking, *The Ethnographer's Magic and Other Essays in the History of Anthropology* (Madison, WI: University of Wisconsin Press, 1992), pp. 12–59.

—, 'Retrospective Prescriptive Reflections', in G. Stocking, *The Ethnographer's Magic and Other Essays in the History of Anthropology* (Madison, WI: University of Wisconsin Press, 1992), pp. 3–11.

—, *After Tylor: British Social Anthropology 1888–1951* (Madison, WI: University of Wisconsin Press, 1995).

—, *Delimiting Anthropology: Occasional Essays and Reflections* (Madison, WI: University of Wisconsin Press, 2001).

Schwartz, J., 'Un Beau Souvenir du Canada', in E. Edwards and J. Hart (eds), *Photographs Objects Histories: On the Materiality of Images* (New York: Routledge, 2004), pp. 16–32.

Taylor, C. L., 'Knox, Robert (1791–1862), Anatomist and Ethnologist', *Oxford Dictionary of National Biography*.

Tedlock, B., 'From Participant Observation to the Observation of Participation: The Emergence of Narrative Ethnography', *Journal of Anthropological Research*, 41 (1991), pp. 61–94.

Thompson, E. P., 'Folklore, Anthropology and Social History', *Indian Historical Review*, 3 (1977), pp. 247–66.

Topham, J., 'Scientific Publishing and the Reading of Science in Nineteenth-Century Britain: A Historiographical Survey and Guide to Sources', *Studies in History and Philosophy of Science*, 31 (2000), pp. 559–612.

Turnbull, C. M., 'Crawfurd, John (1783–1868), Orientalist and Colonial Administrator', *Oxford Dictionary of National Biography*.

Turner, F., 'The Victorian Conflict between Science and Religion: A Professional Dimension', *Isis*, 69 (1978), pp. 356–76.

Tylor, E. B., *Anahuac: or Mexico and the Mexicans, Ancient and Modern* (London: Longman, Green, Longman, and Roberts, 1861).

—, *Researches into the Early History of Mankind and the Development of Civilisation* (London: John Murray, 1865).

—, *Primitive Culture: Researches into the Development of Mythology, Philosophy, Religion, Language, Art and Custom*, 2 vols (London: John Murray, 1871).

—, *Anthropology: An Introduction to the Study of Man and Civilisation* (London: Macmillan, 1881).

—, 'On a Method of Investigating the Development of Institutions; Applied to Laws of Marriage and Descent', *Journal of the Anthropological Institute*, 18 (1888), pp. 245–72.

—, *Primitive Culture: Researches into the Development of Mythology, Philosophy, Religion, Language, Art and Custom*, 2nd edn, 2 vols (London: Henry Holt and Company, 1889).

Urry, J., 'Notes and Queries on Anthropology and the Development of Field Methods in British Anthropology, 1870–1920', *Proceedings of the Royal Anthropological Institute of Great Britain and Ireland*, 1972 (1972), pp. 45–57.

—, 'A History of Field Methods', in R. F. Ellen (ed.), *Ethnographic Research: A Guide to General Conduct* (London: Academic Press, 1984), pp. 27–61.

—, 'Englishmen, Celts and Iberians: The Ethnographic Survey of the United Kingdom', in G Stocking (ed.), *Functionalism Historicised: Essays in British Anthropology* (Madison, WI: University of Wisconsin Press, 1984), pp. 83–105.

—, *Before Social Anthropology: Essays on the History of British Anthropology* (Chur: Harewood Academic Publishers, 1993).

Van Riper, A. B., *Men among the Mammoths* (Chicago, IL: University of Chicago, 1993).

Vesey-FitzGerald, B., *Gypsies of Britain: An Introduction to their History* (Newton Abbot: David and Charles, 1973).

Vogt, C., *Lectures on Man: His Place in Creation, and in the History of the Earth*, ed. J. Hunt (London: Longman, Green, Longman, and Roberts, 1864).

Wallace, A. R., 'On the Origin of Human Races and the Antiquity of Man Deduced from the Theory of Natural Selection', *Journal of the Anthropological Society of London*, 2 (1864), pp. clviii–clxxxvii.

Webster, C., 'The Portrait Cabinet of Dr. Bleek: Anthropometric Photographs by Early Cape Photographers', *Critical Arts*, 14 (2000), pp. 1–15.

White, H., *Tropics of Discourse: Essays in Cultural Criticism* (Baltimore, MD: Johns Hopkins University Press, 1978).

—, *The Content of the Form: Narrative Discourse and Historical Representation* (Baltimore, MD: Johns Hopkins University Press, 1987).

White, P., *Thomas Huxley: Making the 'Man of Science'* (Cambridge: Cambridge University Press, 2003).

Wright, T., 'Preface', *Transactions of the Ethnological Society*, 2 (1863), pp. iii–viii.

Wright, T., 'Photography: Theories of Realism and Convention', in E. Edwards (ed.), *Anthropology and Photography 1860–1920* (New Haven, CT: Yale University Press, 1992), pp. 18–31.

Wright, T., and J. C. Prichard, 'Ethnology', in J. F. W. Herschel (ed.), *A Manual of Scientific Enquiry; Prepared for the Use of Officers in Her Majesty's Navy and Travellers in General* (London: John Murray, 1859), pp. 253–67.

Yeo, R., 'Scientific Method and the Rhetoric of Science in Britain 1830–1917', in J. Schuster and R. Yeo (eds), *The Politics and Rhetoric of Scientific Method* (Dordrecht: D. Reidel Publishing Company, 1986), pp. 259–97.

# INDEX

Printed and bound by CPI Group (UK) Ltd, Croydon, CR0 4YY
16/09/2026
14963484-0001